Springer Theses

Recognizing Outstanding Ph.D. Research

For further volumes:
http://www.springer.com/series/8790

Aims and Scope

The series "Springer Theses" brings together a selection of the very best Ph.D. theses from around the world and across the physical sciences. Nominated and endorsed by two recognized specialists, each published volume has been selected for its scientific excellence and the high impact of its contents for the pertinent field of research. For greater accessibility to non-specialists, the published versions include an extended introduction, as well as a foreword by the student's supervisor explaining the special relevance of the work for the field. As a whole, the series will provide a valuable resource both for newcomers to the research fields described, and for other scientists seeking detailed background information on special questions. Finally, it provides an accredited documentation of the valuable contributions made by today's younger generation of scientists.

Theses are accepted into the series by invited nomination only and must fulfill all of the following criteria

- They must be written in good English.
- The topic should fall within the confines of Chemistry, Physics, Earth Sciences, Engineering and related interdisciplinary fields such as Materials, Nanoscience, Chemical Engineering, Complex Systems and Biophysics.
- The work reported in the thesis must represent a significant scientific advance.
- If the thesis includes previously published material, permission to reproduce this must be gained from the respective copyright holder.
- They must have been examined and passed during the 12 months prior to nomination.
- Each thesis should include a foreword by the supervisor outlining the significance of its content.
- The theses should have a clearly defined structure including an introduction accessible to scientists not expert in that particular field.

Matteo Fumagalli

Increasing Perceptual Skills of Robots Through Proximal Force/Torque Sensors

A Study for the Implementation of Active Compliance on the iCub Humanoid Robot

Doctoral Thesis accepted by
the University of Genoa, Italy

Author
Dr. Matteo Fumagalli
Dipartimento di Informatica, Sistemistica e Telematica
University of Genoa
Genoa
Italy

and

Robotics Brain and Cognitive Science
Italian Institute of Technology
Genoa
Italy

and

Robotics and Mechatronics
University of Twente
Enschede
The Netherlands

Supervisors
Dr. Francesco Nori
Robotics Brain and Cognitive Science
Italian Institute of Technology
Genoa
Italy

Prof. Giorgio Metta
Dipartimento di Informatica, Sistemistica e Telematica
University of Genoa
Genoa
Italy

and

Robotics Brain and Cognitive Science
Italian Institute of Technology
Genoa
Italy

ISSN 2190-5053 ISSN 2190-5061 (electronic)
ISBN 978-3-319-37573-1 ISBN 978-3-319-01122-6 (eBook)
DOI 10.1007/978-3-319-01122-6
Springer Cham Heidelberg New York Dordrecht London

Printed on acid-free paper

Springer is part of Springer Science+Business Media (www.springer.com)

Parts of this thesis have been published in the following articles:

- Fumagalli M., Ivaldi S., Randazzo M., Natale L., Metta G., Sandini G., Nori F., **Force feedback exploiting tactile and proximal force/torque sensing,** (2012) Autonomous Robots, http://www.springerlink.com/index/10:1007/s10514-012-9291-2

- Ivaldi S., Fumagalli M., Randazzo M., Nori F., Metta G., Sandini G., **Computing robot internal/external wrenches by means of inertial, tactile and f/t sensors: theory and implementation on the iCub,** 2011 11th IEEE-RAS International Conference on Humanoid Robots (Humanoids).

- Randazzo M., Fumagalli M., Nori F., Natale L., Metta G., Sandini G., **A comparison between joint level torque sensing and proximal F/T sensor torque estimation: implementation on the iCub,** 2011 IEEE/RSJ International Conference on Intelligent Robots and Systems (IROS), pp: 4161–4167

- Fumagalli M., Randazzo M., Nori F., Natale L., Metta G. and Sandini G. 2010, **Exploiting Proximal F/T Measurements for the iCub Active Compliance,** 2010 IEEE/RSJ International Conference on Intelligent Robots and Systems, Taipei, Taiwan, October 18–22, 2010.

To my family

Supervisor's Foreword

When I first met Matteo, it was autumn 2006. He applied for one of the Ph.D. positions available in the group of Prof. Sandini, Chair of the Robotics Brain and Cognitive Science of the Istituto Italiano di Tecnologia IIT, and assigned by the DIST Department of the University of Genoa. I remember that at that moment, the main structure of the IIT was not yet accessible. Therefore he came to the Liralab, a laboratory in Villa Bonino, nearby the Faculty of Engineering of the University of Genoa. That was the first time I had the pleasure to discuss with him some scientific issues. He looked like the typical student who just achieved his M.Sc. and believed that conquering the world is an easy issue. He showed good attitude in socializing and communicating, but he also gave me the sensation that he was a bit scientifically naive. I must confess that at that moment, although I did not really realize the real potentials of Matteo, he somehow managed to show me something by which I was affected, but that I could not really say what it was.

The second time I met him was already after the selection of the Ph.D. candidates and, at that time, it was possible to access a small part of the building where now the IIT is located. He came and asked what he could do. He wanted to start working on his assignment. I could see that he was really sure of himself, although I knew that the real implementation on robotic systems was not part of his background. Therefore I realize that it was time for me to challenge him. His Ph.D. theme was, in short, the study of force control methods for humanoid robots. Therefore, as first assignment I asked him to implement a software executable to read from a data acquisition board the information of a 6-axis force/torque sensor that was present on the arm of James, a half-torso humanoid robot that we had at the Liralab, and that we moved to the IIT while waiting for the iCub to be fully developed and constructed.

As it is typically the case when you give the first assignment to a just graduated student, the development of this very simple piece of software took longer than expected. Nevertheless, although he spent too much time on it, he gave me a good sensation about one of his skills: he never gives up when trying to solve a problem. Happy with his results, it was time for me to challenge him again. Therefore I asked him to implement a force controller using those measurements. Also, this time it took a long time (not as much as for the first task, but still the response was not immediate) and the results were more than suboptimal. But once again, he did it! The story continued for three more years, and this is how Matteo grew up.

What I saw in Matteos eyes the first time I met him was exactly this! His main characteristic is his willingness to face daily problems and, in one way or another, find a solution. And this solution was presented to me, by him, faster and better. After a while, his results were improving, his outcome started to become interesting. I became interested in what he was doing, to the extent that I decided to finally collaborate with him actively, giving him help also from other resources that I had, such as the implementation skills of other Ph.D. students.

His work not only led to the publication of this very interesting book, but gave origin to a series of research challenges that we are still addressing both in the group and at the European level.

The solutions shown in this thesis allow a complete representation of the interaction forces of the iCub humanoid robot and a nice and cheap way to achieve active compliance. The proposed approach makes use of three sets of sensors, distributed along the kinematic chain: force/torque, inertial, and tactile sensors. The focus of the book is on understanding how to integrate the measurements from these sensors to estimate both internally and externally applied wrenches (i.e. forces and torques). Specifically, the questions addressed in the book will be the following: given a multiple branch kinematic chain with embedded (force/torque, tactile, inertial) sensors, assuming that some unknown wrenches act on the system and assuming that a dynamical model of the system is available:

- is there a systematic procedure to propagate force/torque measurements along the chain in order to estimate internal wrenches? How does this procedure change depending on the location of the externally applied wrenches?
- is there a systematic procedure to measure the external wrenches acting on the chain? How many external wrenches can be estimated? Is there a condition on the localization of these external wrenches (with respect to the FTS locations) to guarantee that they can be estimated?

This book presents a nice mathematical formulation for systematically answering the questions above. The proposed procedure consists of a post-order traversal of a tree, which is obtained by an on-line rearrangement of the graph, according to the contact locations. Conditions on the graph structure will be given in order to guarantee the propagation of force/torque measurements to the entire kinematic chain and, in particular, to the unknowns. It will be shown that given N-FTS, a maximum of N + 1 unknown external wrenches can be estimated, if there is exactly one unknown per each of the subgraphs induced by the force/torque sensors locations. When all external wrenches can be computed, then all joint torques can be computed too.

Genoa, April 2013 Francesco Nori

Preface

This thesis presents the result of 4 years of work as a Ph.D. student at the Department of Informatics, Systems and telecommunication (DIST), University of Genoa, from January 2006 to April 2011. The work has been conducted within the Robotics, Brain and Cognitive Science (RBCS) group of the *Istituto Italiano di Tecnologia*, Genoa (IT), which has hosted me during the entire period of my Ph.D. studies.

When I applied for the Ph.D. position, I proposed to work on interaction control as a tool to create a representation of the environment surrounding the robot. When visiting the group of Prof. Giulio Sandini before the selection of the Ph.D. candidates, I met the person who became my supervisor and friend, Dr. Francesco Nori. He introduced me to the group and showed me the platforms they had. At that moment the iCub robot was still under design process. During the visit, my second advisor Giorgio Metta showed me a presentation about the iCub, giving me an introduction to the research that would have been developed on top of it. In that period, the humanoid platform they were using for doing research was the half-torso called James which became the robot that gave origin to the work that is presented in this dissertation. James is a single arm robot equipped with dexterous hand and head. It mounts a 6-axis force/torque sensor on its arm, between the shoulder and the elbow. The iCub robot has been designed similarly. When describing James during my visit, Francesco said that the reason why the force/torque sensor had been placed in that position was due to the strict design requirements that did not allow to place it *classically* on its wrist.

I think that this manuscript gives a better motivation for this design choice, showing both its advantages and disadvantages compared to other ways to achieve interaction measurements and control.

The dissertation addresses all the people who make use of force/torque sensors to control the interaction of robotic systems. In particular, it deals with the propagation of kinematic and dynamic quantities within the links of the iCub humanoid robot. The goal of this work is to increase the perceptual capabilities of robotic systems and to perform interaction control by means of a limited amount of mechanical sensors. The manuscript shows the insights related to the whole body force control of iCub. It is meant to be both a guide for the iCub and Robotcub communities for understanding the implementation of torque and impedance control on the iCub robot, and also a formal reference for all the roboticians who have a force sensor and don't know where to place it.

In this thesis, the Chap. 1 introduces the issues related to the research on physical interaction and highlights the areas covered by the research reported within this manuscript. Chapter 2 gives a brief overview of the iCub humanoid robot. A description of the main parts constituting the iCub, the sensors it is equipped with, and its electronic and software architecture will be presented. Particular emphasis is given to the description of the sensors that have been fundamental to carry out this work: an inertial sensor placed in the head of the robot, and a total of four force/torque sensors distributed along the kinematic tree, one for each limb, placed in a proximal position.

Chapter 3 introduces the basic concepts of multi-body system dynamics, with particular emphasis on the recursive Newton-Euler algorithm. A graph formulation which uses information on internal sensors will be proposed and presented. This formulation, which has been called *Enhanced Graph Formulation*, allows to easily represent the flow of kinematic and dynamic information along the mechanical structure. The method has the main advantage to exploit force/torque sensor measurements and artificial skin, that allow to represent dynamically the interaction forces that arise on the links, during an interaction scenario, in a non-predetermined position.

Chapter 4 shows a case study of the EOG method presented in Chap. 3 applied to the iCub humanoid robot. The method is validated through experiments which show the effectiveness of the method for internal dynamic estimation, virtual external wrench, and also virtual joint torque measurement.

Chapter 5 shows the study of force control of coupled transmission system. A computed torque approach is proposed for coupled transmissions. On top of it, interaction control strategies are introduced to perform compliant control and joint impedance control of the iCub joints.

Chapter 6 shows the software architecture built for the calculation of the iCub dynamic and the implemented control framework implemented for low level compliance control.

Enschede, April 2013 Matteo Fumagalli

Acknowledgments

I want to thank Prof. Sandini and Prof. Metta for their support and scientific freedom during this 4 years of Ph.D. Their advises have been very helpful.

I want to thank Lorenzo Natale for his helpfulness in discussing every sort of problem.

A particular acknowledge goes to my supervisor and friend, Francesco Nori, whose medieval *actions* aimed at downing me, where able to cause my *reactions* that made me get back on my feet. He is the only one who was able to convince me of my possibilities, not only by saying comfortable words, but by making me face him, myself and reality. I really thank him for his good job with me, even though I am now a fat and shabby nerd.

I give a warm thank to Marco Randazzo, whose collaboration and teaching have been fundamental for my personal growth and Serena, for her collaboration during the conclusion of our Ph.D. period.

Thanks to all the staff of the Robotics Brain and Cognitive Science department of the IIT which helped me with practical issues.

Finally, I thank for the financial support, the Italian Ministry of Education, University and Research (MIUR), the University of Genoa and the Fondazione Istituto Italiano di Tecnologia (IIT).

Contents

Acronyms

DH	Denavit Hartenberg
DoF	Degree of Freedom
DSP	Digital Signal Processor
EOG	Enhanced-Oriented Graph
FTS	Force/Torque Sensor
MBSD	Multi-Body System Dynamics
RNEA	Recursive Newton-Euler Algorithm

Part I
Increasing Perceptual Skills of Robots Through Proximal FTSs

Chapter 1
The Role of Force Perception and Backdrivability in Robot Interaction

As human started to think of robot, safety was already one of the most important issues to achieve. It was the 1942 when Isaac Asimov introduced the laws of robotics in his *Runaround* story.

It was the 1981 when a 37-year old maintenance engineer died in Japan while working to perform maintenance in a restricted safety zone [46].

It is now widely diffused in robotic research the concept that an autonomous robot must be capable to safely interact with the environment and with humans to carry on common duties. Currents trends in robotics boost the research in the development of capabilities and skills which can make robots autonomous and safe (i.e. not dangerous).

These capabilities require behaviors that allow the robot to cope with uncertainties, noises and unpredictable events. One of the main robotic scenarios requires humans and robots to coexist within a shared unstructured environment, to interact and perform both independent and cooperative tasks, precisely and safely.

Depending on the scenario that involves robotic systems, the word safety takes different meanings (see [11, 47] for details about recent trends in robotics).

As an example, fault-tolerant methods (see [25]) have the goal to detect and limit the consequences of hardware and software problems during the robot's motion(see [39]). Danger index can be used to define limitations to the motion of the robot in order to avoid dangerous collisions (see [7, 23]). Collision avoidance solutions monitor the surrounding of the robot to generate trajectories that avoid collisions with the external elements or the robot's own structure and still represent an open issue in robotics (see [22, 24, 30, 40]).

More in general, safety related issues of an autonomous system rely on its sensor system. Therefore it is of primary importance to provide the robot with information that allow to correctly represent the situation and the interaction. By means of the sensors that constitute the perceptual capabilities of the robot, the autonomous system creates its own knowledge of the surrounding.

The sensor system of robots can in general be divided into two main categories: the *high-level sensor system* and the *low level sensor system*. The former allows to obtain

M. Fumagalli, *Increasing Perceptual Skills of Robots Through Proximal Force/Torque Sensors*, Springer Theses, DOI: 10.1007/978-3-319-01122-6_1,

a wide representation of the surrounding environment. Their measurement is very informative, but the interpretation is complicated and computationally expensive. Belonging to this category there are cameras, laser, proximity sensors, radar. On the other hand, the low level sensor system gives a detailed information about the state of the robot. This information is typically localized and can be easily processed. Joint torque sensors and encoders are classical example of this type of sensor system.

In general, all the possible sources of information (cameras, proximity sensors, FTSs, proprioceptive sensors, etc.) must be included to improve the representation of the robot workspace, in order to fulfill the autonomous tasks while preserving safety. It must be noted that achieving safety is not only a matter of properly represent the surrounding and the interaction and control it by means of feedback loops. Mechanical choices can additionally reduce the consequences of failures in the representation of the interaction.

Hereafter, a brief analysis describing the main solutions that are object of studies in the field of Physical Human Robot Interaction (PHRI) is reported. These solution include mechanical choices, ways of measuring the interaction and control solutions that are typically adopted to prevent unsafe interaction and to control the robot behavior.

1.1 State of the Art on Physical Human Robot Interaction: Mechanical Solutions

The main subjects of study in the field of PHRI can be divided into two main categories: those implemented into mechanical solutions and that intrinsically guarantee a certain level of safety, and those requiring sensors measurements and control.

1.1.1 Mechanical Solutions

A wide branch of the research on robot interaction and robot safety focuses on the definition and implementation of innovative mechanical solutions that allow to limit the risk of damage and injuries due to the robot interaction. This is the case, for example, of *lightweight design*, *backdrivable systems* and *compliant mechanisms*.

Light Weight Design

It has been shown in [1], [11], [14] and [15] that among the possible criterion that limit injuries that can be caused by collisions, weight reduction and limited inertia of the moving parts is fundamental. In [16] an evaluation of the damages that can be caused to a (dummy) human hit by a light weight robot is shown. It has been shown that lightweight robot have the intrinsic capability to reduce damage caused by the moving parts. Nevertheless, the typically high reduction ratio of the gearboxes introduces limitations to light weight designs. On one hand, in fact, it allows to reduce the dimensions of the motor. On the other hand, the use of gearboxes introduces the problematic of the reflected inertia. As a consequence, the overall inertia of the load side is:

$$J_{all} = n^2 J_m + J_l \tag{1.1}$$

One effective solution to reduce the inertia of the moving parts is to use cable transmissions. Motors are typically the major source of weight for robots moved by electro-magnetic actuators. One possibility, whose major example is constituted by the Barret Whole Arm Manipulator (WAM) [4, 33, 42], is to place the motors at the base of the robot and transfer the power to the joints by means of cable transmissions. Finally, in order to increase the overall motor power in a larger bandwidth, solutions such as the *Macro-Mini* actuation (see [48]) have been developed. Here, a powerful motor which guarantee low bandwidth power generation is placed on the base of the links, while a smaller motor is placed on the joint and provides torque for high frequency tasks.

Backdrivable Systems

In order to avoid the use of gearboxes or complex mechanical solutions, backdrivable systems can represent an interesting alternative. The term *backdrivability* refers to the easiness of transmission of movement from the output axis, to the input axis, as a consequence of an externally applied force. Different definition have been given in literature by [20, 33, 42].

Compared to non backdrivable systems, this solution generally presents higher bandwidth of the interaction control loop, which reduces the limitations related to passivity related issues of non backdrivable systems, as shown in [9, 19].

Compliant Mechanisms

Another possibility to reduce injury risks in physical human robot interaction is to add compliance. Compliance can be added to the exterior part of the robot (e.g. soft covers) to limit the effect of impact with rigid surfaces, but this aspect does not solve the issue of reducing the problems related with the reflected inertia at impacts.

One solution that allows to mechanically decouple the large reflected inertia of the motors from those of the links is to add elastic elements in the joint transmission. Compliant transmission may ensure safe interaction, since the intrinsic elasticity stores the energy of impacts into the springs, therefore limiting the energy lost in the impact. These elements in fact allow to convert the kinetic energy of the moving links into potential energy of springs. Example of such systems are the *Series Elastic Actuators* (SEA), designed at MIT by Pratt and Williamson, which proposed the first prototype in 1995 [32]. The SEA are characterized by an elastic interconnection element between the motor and the load. SEA have been employed for robotic mechanisms such as the humanoid robots COG [3], DOMO [12] and TWENDY-ONE [21].

A more recent solution to the problem related with SEA, is to give to the system the possibility to vary its own internal joint stiffness. This is the case of Variable Impedance Actuators (VIA) (see [2, 35, 41]), where the stiffness of the joint can be mechanically varied in order to have improved motion performances while moving (high stiffness), and safe interaction when contact occurs (low stiffness). On the other hand, recent studies have shown that certain solutions present an undesired effect that may lead to unsafe behaviors. Elastic elements, especially if combined with actuators can store great amounts of potential energy which, once released, can be extremely unsafe, as recently shown in [17].

1.2 Measurements of the Interaction and Control

Another way to obtain safe behavior of the robotic platform, is by means of control solutions. When *active compliance* is employed, the robot behavior relies on the sensory system. Force information are necessary for the autonomous system to build its own *epistemology* of the interaction. Exploiting these information the robot can carry out force regulation [8, 37], react to contacts and also take decisions about the tasks (see [10, 13, 18, 31]). Measurements about the interaction, for control, are typically retrieved in different ways. Current measurements, joint torque sensors or 6-axis force/torque sensors (the latter typically placed at the end-effector), are the most commonly used measurements of the interaction of the system.

Current Measurements

Current measurement of electric motors can be exploited to have an estimation of the torque that the motor transmits to the load through the electro-mechanical coupling:

$$\begin{cases} I\ddot{\theta} + f_c + f_d = \tau_m - \tau_l \\ L\frac{d}{dt}i + Ri - K\dot{\theta} = V \\ \tau_m = Ki \end{cases} \tag{1.2}$$

If the mechanical part presents sources of dissipation (e.g. in the motor or in the transmission, such as coulomb friction f_c and dynamic friction f_d), the actual torque that is transmitted to the load is reduced by these factors. It is remarkable that the lower the reduction ratio, the more effective is the approach. Therefore low reduction gearboxes with high efficiency are typically employed when current measurements are performed. This allows the design of system with relatively small motors (lightweight), still maintaining high torque transmission, but with the advantage that the reflected inertia is small. This solution have been used in fact for the whole arm manipulator (WAM) proposed by [33].

On the other hand, high reduction ratios are typically inefficient and introduce dissipation and frictional related problems, thus influencing the measurement of the current. Therefore, current measurements cannot be used to properly represent the torque that is acting at the load side.

1.2.1 Joint Torque Sensing

Joint torque sensors are a possible solution to the problems that remain unsolved given the approaches previously mentioned.

Torque sensors are typically placed on the load side of the joint. They allow to have a reliable information of the interaction that occur, without being influenced by problems related to the friction and inefficiency of the gearboxes. Moreover they are distributed along the kinematic chain, which is advantageous for measuring the interaction at various points of the robotic structure. An exhaustive study about explicit force control have been conducted in [45], where different control strategies and stability analysis have been reported.

On the other hand, they do not allow to obtain a complete representation of the interaction. Joint level torque sensors measure the torque that is working along the joint axis. Therefore, the retrieved information is configuration dependent, and it is not guaranteed that a complete representation of the interaction is obtained by means of joint torque sensors.

1.2.2 Force Measurements

The information of 6-axis force/torque sensors give the most complete knowledge of the interaction. Force sensors in fact directly measure the generalized force that is applied at the sensor frame. These sensor have been widely employed for the control of the interaction in manipulation tasks. They have been classically employed for research in the control of the interaction of industrial robots (see [6, 36, 38]).

Classical applications place this powerful source of information at the end-effector of the manipulator structure. The reason for this, is that we are interested in retrieving a direct measurement of such information at the point that we want the interaction to occur. Given the measurement at the tool level, we can think of projecting it at the joint level through the transpose of the Jacobian of the manipulator, and thus perform joint level torque control.

$$\tau = J^{\top} F \tag{1.3}$$

Nevertheless, this limitation might be valid in an industrial scenario. When the robot is working in an unstructured environment, the possible point of interaction is not known a priori. Other sensors can work together to prevent interaction at other levels (such as cameras or proximity sensors), but they cannot give any sort of representation of the actual interaction and transmission of generalized force between the robot and the object or human. In this situation a different framework is required. A method to obtain a more distributed measurement of the interaction is necessary.

1.3 Significance of the Thesis

Recent robotic trends and future perspective of robotics are defining new guidelines of robotic research. Autonomous robots should be capable of coexisting with humans in an environment which is dynamically changeable. Their skills and behaviors should be dominated by processes that are capable of adapting to different situation and unpredictable events. Their actions should be the consequences of events, but events should also be the cause of the adaptation process that allow the autonomous system to create its knowledge and representation of the action and reaction itself.

One of the possible ways of organizing the knowledge of the autonomous system is by means of the process of *enaction*. Enaction can be defined as a form of knowledge that starts from the interaction with the world. A first definition of this term have been proposed in [5], and has been refined by Varela and Maturana (see [26–28, 44]), who stated that the enactive knowledge is a particular type of knowledge that is constructed on the motor skills required to perform an action such as manipulating objects, riding a bicycle or playing a sport. Therefore, the enactive knowledge can be considered as that type of knowledge that is acquired by doing an action.

In [34] it is stated that the developmental cognitive architecture must me capable of adaptation and self-modification. Therefore, the autonomous system must be able to adapt and modify the parameters which determine its phylogenetic skills by doing experience. By means of a learning process, it should be able to modify its own cognitive structure and organization. Through the adaptation based on experience, it should be able to modify its system dynamics, enlarge its repertoire of actions and adapt to novel circumstances (see [34]).

Therefore, it is important that the development of the autonomous system is based on explorative behaviors. Only with force exploration, and therefore by the acquisition of haptic and force information, the cognitive system can experience and adapt to the physical world and, by collecting these information, it can create its own representation of the surrounding [29].

Cognitive processes have been classically studied as abstract theories, mathematical models, and disembodied artificial intelligence. Nevertheless, the cognitive processes are strongly entwined with the physical structure of the body and its capability to interact with the environment.

Therefore, intelligence and mental processes are deeply influenced by the structure of the body, by its motor capabilities and by the morphology of the sensory system. The way the physical body and its actions influence the development of the autonomous system is not less important that the one that neural processes have.

Human intelligence develops indeed through interaction with objects in the environment, and is also strictly connected to the interactions with other autonomous systems.

In other words, the cognitive and physical systems are not independent: on one hand, physical interaction sets the rules for the cognitive evaluations of the environment during interaction tasks; on the other hand, cognitive aspects improve the way the system interacts by setting suitable control interaction parameters. Therefore, force perception plays a fundamental role for the cognitive development of robotic systems. Since it is impossible to model every action in an unstructured anthropic environment, the intelligent connection of perception with action of robots implies the presence of autonomous behavior, and therefore an interconnection of sensing capabilities and cognitive processes, to solve real problems.

This work is aimed by the need of distributed force perception for the definition of a framework in which the iCub humanoid robot [43] perceives both external and internal forces during the exploration process.

Giving the assumptions previously shown, stating that robot interaction is at the basis of cognitive processes (a more accurate discussion can be found in [34]), the cognitive systems creates in this way its own knowledge (epistemology) of the surrounding environment.

Therefore, the robot should be equipped with low level basic behaviors that allow to obtain a complete perception of the interaction, in order to control it in a continuously evolving environment.

By exploiting force information, proper control strategies can be adopted. Active compliance allows to reduce, at certain frequencies, the admittance of the system, therefore resulting in a more backdrivable and safe behavior of the robot.

However, the classical ways presented in Sect. 1.2, which employs localized sensors, do not generally allow a full perceptual representation of the interaction scenario in terms of forces and torques which rise over the whole structure.

More precisely, although the use of joint torque sensors (or current sensors) guarantees a good approach for distributed sensing over the entire structure of the robot, their measurements lack of completeness and, depending on the configuration of the robot, they suffer from null projection on the joint angles of the externally applied forces.

This work focuses on the development of methodologies that enrich the perceptual capabilities of the interaction of the iCub robot by exploiting a distributed set of FTSs over the robot's structure. The method have been implemented to give to the robot the possibility to perceive the interaction wherever it occurs and, by means of control solution, react to it. This work is fundamental for the creation of a basic and distributed low level sensor system, necessary to perform the exploration and to create a proper representation of the surrounding of robotic systems.

References

1. Alami, R., Albu-Schaeffer, A., Bicchi, A., Bischoff, R., Chatila, R., Luca, A. D., et al. (2006). Safe and dependable physical human-robot interaction in anthropic domains: state of the art and challenges. In A. Bicchi & A. D. Luca (Eds.), *Proceedings IROS Workshop on pHRI—Physical Human-Robot Interaction in Anthropic Domains*, Beijing, China.
2. Bicchi, A., & Tonietti, G. (2004). Fast and "soft-arm" tactics [robot arm design]. *Robotics Automation Magazine, IEEE, 11*, 22–33.
3. Brooks, R.A., Breazeal, C., Marjanovic, M., Scassellati, B., & Williamson, M.M. (1999). The Cog project: Building a humanoid robot. In C. L. Nehaniv (Ed.), *Computation for metaphors, analogy, and agents*. Lecture notes in computer science (pp. 52–87). Berlin: Springer.
4. B.T. Inc. (2010). *The WAM arm from Barret Technology*. http://www.barrett.com/robot/products-arm.htm.
5. Bruner, J. S. (1968). *The Processes of cognitive growth: Infancy*. Worcester, MA: Clark University Press.
6. Caccavale, F., Natale, C., Siciliano, B., & Villani, L. (2005). Integration for the next generation: Embedding force control into industrial robots. *Robotics Automation Magazine, IEEE, 12*, 53–64.
7. Calinon, S., Sardellitti, I., & Caldwell, D. (2010). Learning-based control strategy for safe human-robot interaction exploiting task and robot redundancies. In *IEEE/RSJ International Conference on Intelligent Robots and Systems*, Taipei, Taiwan.
8. Chiaverini, S., Siciliano, B., & Villani, L. (1999). A survey of robot interaction control schemes with experimental comparison. *IEEE/ASME Transactions on Mechatronics, 4*, 273–285.
9. Colgate, J.E. (1988). *The control of dynamically interacting systems*. Thesis (Ph. D.), Massachusetts Institute of Technology, Cambridge, MA, USA.
10. De Luca, A. (2006). Collision detection and safe reaction with the DLR-III lightweight manipulator arm. In *IEEE/RSJ International Conference on Intelligent Robots and Systems* (pp. 1623–1630).
11. Desantis, A., Siciliano, B., Deluca, A., & Bicchi, A. (2008). An atlas of physical human–robot interaction. *Mechanism and Machine Theory, 43*, 253–270.
12. Edsinger-Gonzales, A., & Weber, J. (2004). Domo: A force sensing humanoid robot for manipulation research. *Humanoid Robots, 2004 4th IEEE/RAS International Conference on* (Vol. 1, pp. 273–291).
13. Fumagalli, M., Randazzo, M., Nori, F., Natale, L., Metta, G., & Sandini, G. (2010). Exploiting proximal F/T measurements for the iCub active compliance. In *IEEE/RSJ International Conference on Intelligent Robots and Systems*, Taipei, Taiwan.
14. Haddadin, S., Albu-Schaffer, A., & Hirzinger, G. (2008). The role of the robot mass and velocity in physical human-robot interaction—Part I: Non-constrained blunt impacts. In *IEEE International Conference on Robotics and Automation*, Pasadena, CA, USA.
15. Haddadin, S., Albu-Schaffer, A., Frommberger, M., & Hirzinger, G. (2008). The role of the robot mass and velocity in physical human-robot interaction—Part II: Constrained blunt impacts. In *IEEE International Conference on Robotics and Automation*, Pasadena, CA, USA.
16. Haddadin, S., Albu-Schaffer, A., & Hirzinger, G. (2007). Safety evaluation of physical human-robot interaction via crash-testing. In *Robotics: Science and System Conference (RSS 2007)*, Atlanta, Georgia.
17. Haddadin, S., Albu-Schaffer, A., Eiberger, O., & Hirzinger, G. (2010). New insights concerning intrinsic joint elasticity for safety. In *IEEE/RSJ International Conference on Intelligent Robots and Systems*.
18. Haddadin, S., Urbanek, H., Parusel, S., Burschka, D., Rossmann, J., Albu-Schaffer, A., et al. (2010). Real-time reactive motion generation based on variable attractor dynamics and shaped velocities. In *Intelligent Robots and Systems (IROS), 2010 IEEE/RSJ International Conference on* (pp. 3109–3116).
19. Hogan, N., & Buerger, S. (2004). *Impedance and interaction control* (Chap. 19, pp. 1–23). Boca Raton: CRC Press.

20. Ishida, T., & Takanishi, A. (2006). A robot actuator development with high backdrivability. In *Robotics, Automation and Mechatronics, 2006 IEEE Conference on* (pp. 1–6).
21. Iwata, H., & Sugano, S. (2009). Design of human symbiotic robot TWENDY-ONE. In *2009 IEEE International Conference on Robotics and Automation* (pp. 580–586), IEEE.
22. Khatib, O. (1985). Real-time obstacle avoidance for manipulators and mobile robots. In *Robotics and Automation. Proceedings. 1985 IEEE International Conference on* (Vol. 2, pp. 500–505).
23. Kulic, D., & Croft, E. (2005). Real-time safety for human–robot interaction. In *Advanced Robotics, 2005. ICAR '05. Proceedings., 12th International Conference on* (pp. 719–724).
24. Kulic, D., & Croft, E. (2007). Pre-collision safety strategies for human-robot interaction. *Autonomous Robots*, *22*, 149–164.
25. Lussier, B., Chatila, R., Guiochet, J., Ingr, F., Lampe, R., olivier Killijian, M. et al. (2005). Fault tolerance in autonomous systems: How and how much. In *Proceedings of the 4th IARP/IEEE-RAS/EURON Joint Workshop on Technical Challenge for Dependable Robots in Human Environments* (pp. 16–18).
26. Maturana, H. (1970). *Biology of cognition*. Research Report BCL 9.0, University of Illinois, Urbana, IL.
27. Maturana, H. (1975). The organization of the living: A theory of the living organization. *The International Journal of Man-Machine Studies, 7*, pp. 313–332.
28. Maturana, H. R., & Varela, F. J. (1980). *Autopoiesis and cognition: The realization of the living*. Dordrecht: D. Reidel Publishing Company.
29. Metta, G., Natale, L., Nori, F., Sandini, G., Vernon, D., Fadiga, L., et al. (2010). The iCub humanoid robot: An open-systems platform for research in cognitive development. *Neural Networks, Special Issue on Social Cognition: From Babies to Robots*, *23*, 1125–1134.
30. Minguez, J., Lamiraux, F., & Laumond, J. P. (2008). Motion planning and obstacle avoidance. In B. Siciliano & O. Khatib (Eds.), *Springer handbook of robotics* (pp. 827–852). Berlin, Germany: Springer.
31. Mistry, M., Buchli, J., & Schaal, S. (2010). Inverse dynamics control of floating base systems using orthogonal decomposition. In *IEEE International Conference on Robotics and Automation*.
32. Pratt, G., & Williamson, M. (1995). Series elastic actuators. In *IEEE/RSJ International Conference on Intelligent Robots and Systems* (pp. 399–406), Los Alamitos, CA, USA.
33. Salisbury, K., Townsend, W., Ebrman, B., & DiPietro, D. (1988). Preliminary design of a whole-arm manipulation system (WAMS). In *Robotics and Automation, 1988. Proceedings, 1988 IEEE International Conference on* (Vol. 1, pp. 254–260).
34. Sandini, G., Metta, G., & Vernon, D. (2007). The iCub cognitive humanoid robot: An open-system research platform for enactive cognition. 50 Years of AI. *LNAI*, *4850*, 359–370.
35. Schiavi, R., Grioli, G., Sen, S., & Bicchi, A.(2008). VSA-II: A novel prototype of variable stiffness actuator for safe and performing robots interacting with humans. In *2008 IEEE International Conference on Robotics and Automation*, Pasadena, CA, USA.
36. Sciavicco, L., & Siciliano, B. (2005). *Modelling and control of robot manipulators*. Advanced textbooks in control and signal processing. Berlin: Springer.
37. Siciliano, B., & Villani, L. (1996). A passivity-based approach to force regulation and motion control of robot manipulators. *Automatica*, *32*, 443–447.
38. Siciliano, B., & Villani, L. (2000). *Robot force control*. Norwell, MA: Kluwer Academic Publishers.
39. Sisbot, E., Marin, L., Alami, R., & Simeon, T. (2006). A mobile robot that performs human acceptable motions. In *Intelligent Robots and Systems, 2006 IEEE/RSJ International Conference on* (pp. 1811–1816).
40. Sisbot, E., Marin-Urias, L., Broqure, X., Sidobre, D., & Alami, R. (2010). Synthesizing robot motions adapted to human presence. *International Journal of Social Robotics*, *2*, 329–343.
41. Tonietti, G., Schiavi, R., & Bicchi, A. (2005). Design and control of a variable stiffness actuator for safe and fast physical human/robot interaction. In *Proceedings of the 2005 IEEE International Conference on Robotics and Automation, 2005. ICRA 2005.*

42. Townsend, W. T. (1988). *The effect of transmission design on force-controlled manipulator performance*. Cambridge, MA: Massachusetts Institute of Technology.
43. Tsagarakis, N., Metta, G., Sandini, G., Vernon, D., Beira, R., Santos-Victor, J., et al. (2007). iCub—the design and realization of an open humanoid platform for cognitive and neuroscience research. *International Journal of Advanced Robotics, 21*(10), 1151–1175.
44. Varela, F. (1979). *Principles of biological autonomy*. New York: Elsevier North Holland.
45. Volpe, R., & Khosla, P. (1993, November). A theoretical and experimental investigation of explicit force control strategies for manipulators. *IEEE Transactions on Automation Control, 30*(11), 1634–1650.
46. Weng, Y. H., Chen, C. H., & Sun, C. T. (2009). Toward the human–robot co-existence society: On safety intelligence for next generation robots. *International Journal of Social Robotics, 1*, 267–282. doi:10.1007/s12369-009-0019-1.
47. Zinn, M., Khatib, O., Roth, B., & Salisbury, J. (2004). Playing it safe—human-friendly robots. *Robotics & Automation Magazine, IEEE, 11*, 12–21.
48. Zinn, M., Khatib, O., Roth, B., & Salisbury, J. (2002). A new actuation approach for human friendly robot design. In B. Siciliano & E. P. Dario (Eds.), *Springer tracts in advanced robotics* (Vol. VIII). Berlin: Springer.

Chapter 2
Platform

Abstract The iCub robot is a humanoid robot that have been developed at the Italian Institute of technology and University of Genoa, with the purpose to carry out research in embodied cognition. More precisely, the iCub robot is the result of a research project focusing on the study of developmental capabilities of cognitive systems. This project, called *RobotCub*, has been funded by the European Commission through Unit E5 *Cognitive Systems, Interaction & Robotics* (see [6, 9, 10, 15, 16]). In the framework of embodied cognition, perception covers a fundamental role in robot learning and development. They are in fact the result of the continuous interaction with the environment, which is necessary to build the basis to abstract reasoning. To achieve learning capabilities and self development and reasoning, *anthropomorphism*, *compliance* and *sensorization* are fundamental. These aspects are necessary to study human and humanoid development. To allow the robot to become a self-reasoning system, the basic perceptual aspects of human become requisites of the design process of a humanoid platform. The main features characterizing the design of the iCub robot, including the choice of the sensors the robotic platform is equipped with, are therefore here presented.

2.1 The iCub Platform

This Section reports an overview of the design of iCub humanoid robot (Fig. 2.1). The goal is to show the *sensori-motor system* characterizing the iCub humanoid robot and some design choices that have been taken during its mechatronic design. The informations that will be presented in this chapter are preparatory to fully understand the goal of this thesis. The following sections focus on introducing the motion capabilities of the iCub robot, its motors and sensors placement. It also gives a preliminary overview of the electronic and software components (which instead will be described in detail in Chap. 6).

M. Fumagalli, *Increasing Perceptual Skills of Robots Through Proximal Force/Torque Sensors*, Springer Theses, DOI: 10.1007/978-3-319-01122-6_2,

2.1.1 Overview of the iCub Robot

The iCub humanoid robot has been designed with the goal of creating an open hardware/software robotic platform for research in embodied cognition, as expressed in [15]. Its design has been mainly developed within RobotCub,[1] a European funded project focusing on the study of natural and artificial cognitive systems (see [6]). One of the major design specification is that the iCub robot should be capable of interacting with humans and environments, using its sensors to react as a response to external events. At the current state, the iCub robot has a total of 53 degrees of freedom (DOF), 6 in each leg, 7 in each arm, 6 in the head, 3 in the waist and 9 for each hand. The iCub employs brushless and brushed DC motors equipped with high reduction gearboxes. In order to both increase the torque capability of the electric motors actuating the robot's joints, and to optimize the compactness of the mechanical design of the robotic platform iself, *harmonic drivers* are mainly employed with brush-less DC motors (1:100 or bigger reduction), while planetary gearboxes engage the brushed DC motors (with reduction ratio spanning from 1:256 to 1:1024). Brushless motors have been employed in the bigger articulated joints (shoulders, elbows, hips, torso, knees) while small brushed motors actuate the distal degrees of freedom (hand joints, neck, eyes). Non standard mechanical choices have been used to rise the torque/volume ratio as will be briefly shown in next subsections. Particular attention will be given to the mechanisms characterizing the shoulder of the iCub robot (see Sect. 2.1.1.1 and also Chap. 5). It must be noted that, although the choice of this type of actuation system allows obtaining an extremely compact designs of the mechanics of the system, it also makes the robot passively non back-drivable.

2.1.1.1 Arm

The iCub arms are 7 DOF open kinematic chains (Fig. 2.2). Their upper part is commanded by four brushless motors, three for the shoulder movements and one for the elbow. The shoulder joint is a 3 DoF cable differential mechanism exploiting a coupled differential transmission system (see Fig. 2.3). Three parallel motors (brushless frameless motors, RBE Kollmorgen series, with harmonic drive reductions, CSD series with 100:1 ratio) housed in the upper-torso move pulleys to generate the spherical motion of the shoulder (see [15, 8] for a more detailed description of the shoulder universal joint). Compared to serial solutions (see e.g. the HRP-2 arm [3]) where every motor actuates the corresponding DoF, the solution adopted for the iCub shoulder engages one motor (Motor 1 in Fig. 2.3) to directly actuate the shoulder pitch, while two additional motors (namely Motor 2 and Motor 3 of Fig. 2.3) actuate two pulleys that are parallel to Motor 1. Motor 1 can deliver up to 40 Nm at the output shaft (after the gearbox), while the other two motors can provide a maximum torque of 20 Nm. The transmission of the motor movements to joint movements is achieved

[1] RobotCub project IST-FP6-004370.

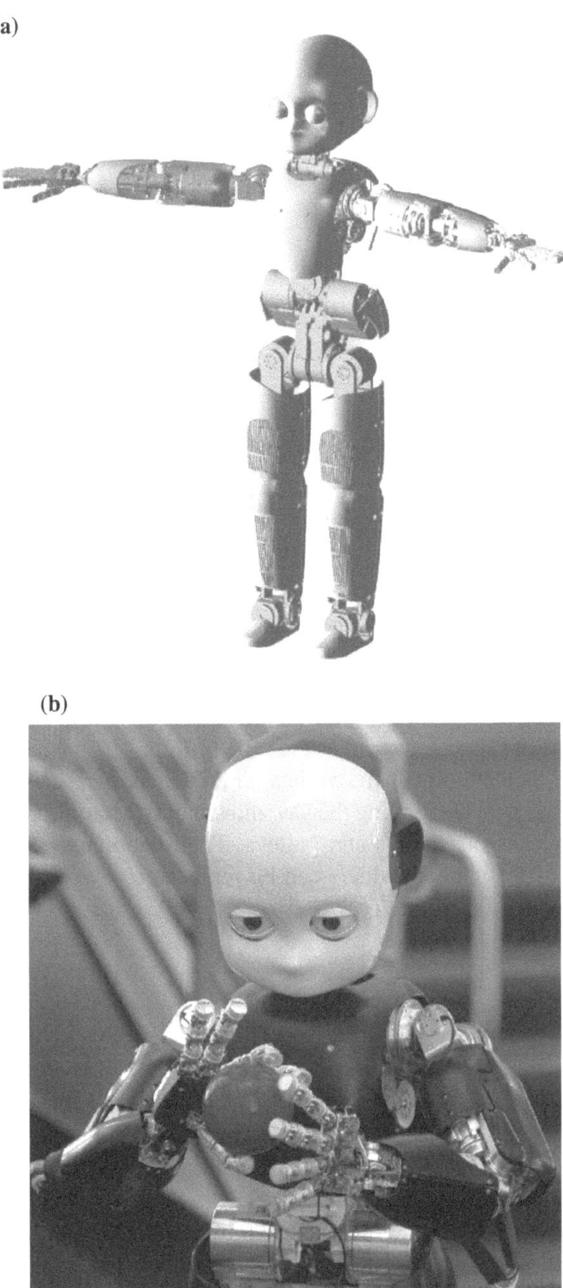

Fig. 2.1 **a** A CAD view of the iCub humanoid robot. **b** A picture of the icub manipulating an object

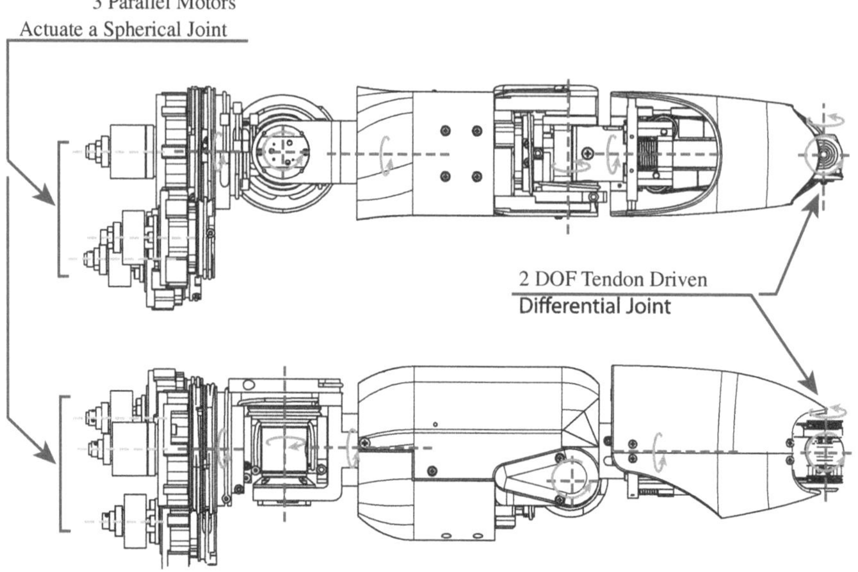

Fig. 2.2 Sketch of the 7 degrees of freedom arm of iCub

by means of idle pulleys. Joint positions are measured by Hall effect based digital encoders with custom made electronics (see [4]). Motor 1 and Motor 3 mount the encoders directly on the motor shaft, while a third encoder measures the position of Joint 2 (actually the joint performing the yaw movement). Motors and joint variables are coupled by means of tendons and pulleys. The relationhip between joint and motor variables of the iCub shoulders can be represented by means of a constant transformation matrix T_{mj} of the form:

$$\dot{\theta}_m = T_{mj}\dot{\theta}_j \qquad T_{mj} = \begin{bmatrix} 1 & 0 & 0 \\ -r & r & 0 \\ -2r & r & r \end{bmatrix}, \tag{2.1}$$

where r is a constant value which depends on the radius of the pulleys, $\dot{\theta}_m = [\dot{\theta}_{m1}, \dot{\theta}_{m2}, \dot{\theta}_{m3}]^\top$ is the vector of motor angular velocities and $\dot{\theta}_j = [\dot{\theta}_{pitch}, \dot{\theta}_{roll}, \dot{\theta}_{yaw}]^\top$ is the vector of joint angular velocities. The generalized dynamic of coupled system will be analyzed in Chap. 5 (Figs. 2.4, 2.5).

The elbow flexion/extension joint is actuated by an independent frameless brushless motor located in the upper arm. The joint is commanded by means of tendons in push-pull configuration, moving an idle pulley. The wrist is designed to be a 3 DOFs spherical joint, allowing pitch, roll and yaw rotations (θ_{wp}, θ_{wr}, θ_{wy}), which correspond respectively to wrist flexion/extension, adduction/abduction and rotation. The roll movement is performed by a single brushed motor directly coupled to the forearm. The pitch and yaw movements instead are accomplished by two parallel

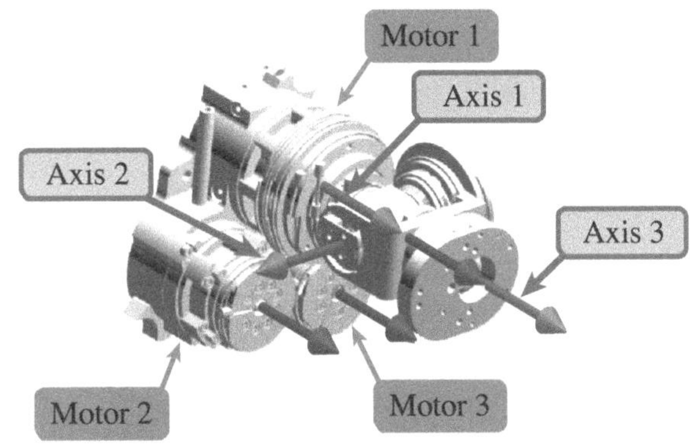

Fig. 2.3 The iCub shoulder. A CAD view of the shoulder joint mechanism showing the *three motors* actuating the 3 degrees of freedom universal joint

motors actuating a semi-differential tendon driven mechanism. Brushed DC motors are exploited for motion the wrist joints. Positions are measured through magnetic incremental encoders mounted on the motors.

2.1.1.2 Hand

The iCub hand (see Fig. 2.4) has five fingers actuated by means of tendon driven mechanisms. Each finger has four joints. The overall number of joints is 17, while number of DoF is 9. Seven motors are placed remotely in the forearm and all tendons are routed through the wrist mechanism. Two only motors are mounted directly on the hand: one of the motors controls the thumb abduction and the other actuates the fingers adduction/abduction. This solution allows locating most of the hand weight in the forearm, resulting in a compact and lightweight hand mechanism. Each joint mounts a tiny Hall effect position sensors. Finger positions are measured by means o a total of 15 Hall effect sensors directly mounted on the phalanxes (three for each finger).

The resulting dimensions are as follows: the palm is 50 mm long, 34 mm wide at the wrist and 60 mm wide at the fingers. The overall thickness of the hand is only 25 mm.

Additionally on the palm, a tactile array is mounted. This tactile sensor is a capacitance based touch sensor, which employs soft silicon rubber as dielectric material (see [1, 5]). On the tip of the fingers, a similar sensor is present, one for each finger (see [13]).

2.1.1.3 Head and Waist

The head (Fig. 2.5) is equipped with two eyes capable to pan and tilt independently (4 DOFs). It is mounted on a 3 DoF neck, which allows the rotation of the head (pitch, roll and yaw movements). Three brushed DC motors actuate the eyes control

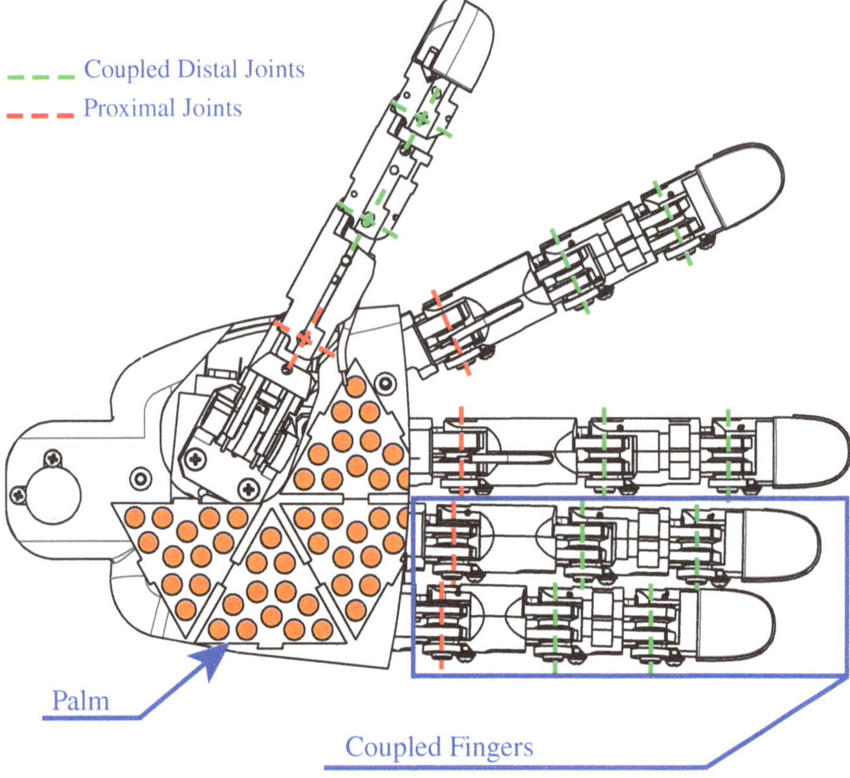

Fig. 2.4 Particular CAD view of the iCub hand. 21 joints are actuated with 9 motors through tendon driven mechanisms. Tactile sensors are present on the palm

independently the pan and simultaneously the tilt. Each eye has a camera which provide the iCub with a vision system. The neck engages a 3 DoF serial solution and is actuated by means of brushed DC motors and planetary gearboxes. The head mounts a PC104, which represents the core of the iCub motion control system. It communicates with other computers and the user by means of Ethernet BUS, as will be deeply described in Sect. 2.2.2.

The waist mechanism (Fig. 2.6) is a 3DOF kinematic chain (see [16, 18]) which makes use of a tendon-driven differential mechanism for pitch and yaw movements, exploiting two parallel brushless motors. This configuration allows a better distribution of the joint torque on the motors, with the advantage of reducing the mechanism dimension. Similarly to the shoulder actuation, the motor velocities $\dot{\theta}_m = [\dot{\theta}_{m1}, \dot{\theta}_{m2}]^\top$ and the joint velocities $\dot{\theta}_j = [\dot{\theta}_{pitch}, \dot{\theta}_{yaw}]^\top$ are kinematically coupled by the linear relation:

$$\dot{\theta}_m = T_w \dot{\theta}_j \qquad T_w = \begin{bmatrix} d & d \\ -d & d \end{bmatrix}, \tag{2.2}$$

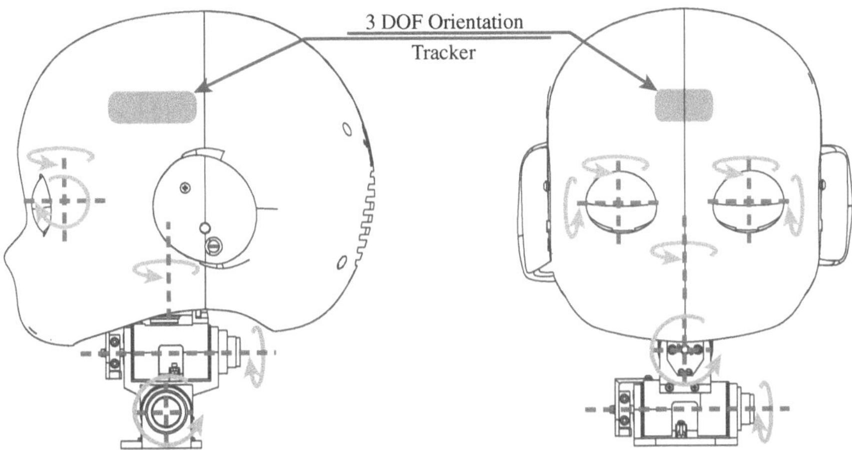

Fig. 2.5 The 7 DOF head of the iCub robot. A 3 DOF Orientation Tracker is mounted at the *top* of the head

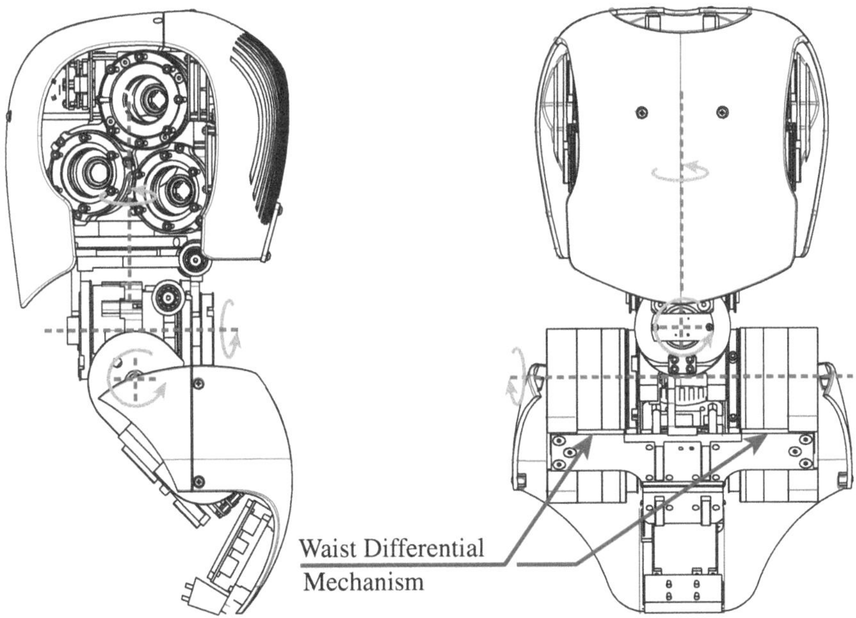

Fig. 2.6 Sketch of the 3 DOF torso. A particular differential mechanism allow to rise the torque to volume ratio of the motors

where d is a constant value which depend on the dimension of the pulleys. The roll movement is actuated independently by a frameless brush-less DC motor with harmonic drive reduction (1:100).

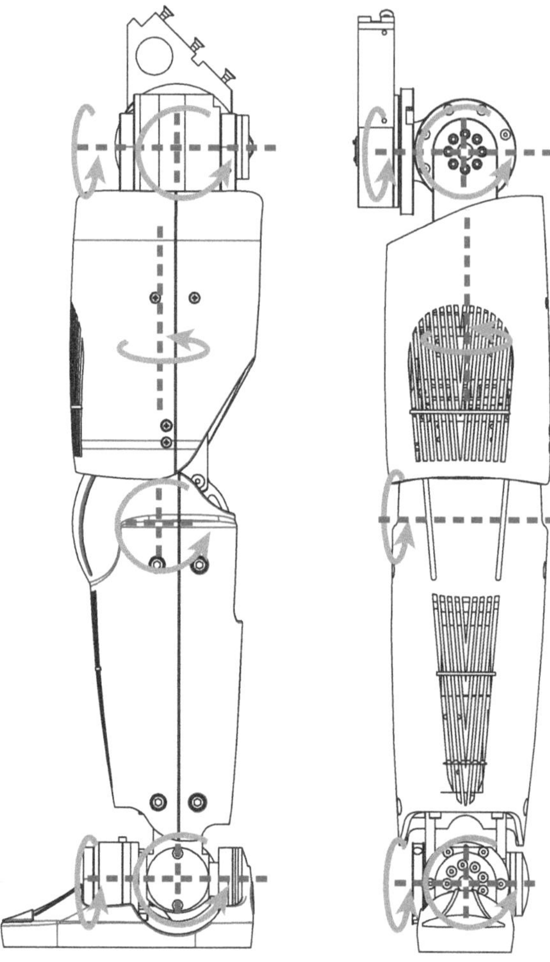

Fig. 2.7 The 6 DOF leg of the iCub humanoid robots

2.1.1.4 Legs

Legs are 6DOF serial kinematic (see Fig. 2.7). A detailed description of these mechanisms can be found in [16, 18]. All the six motors are frameless brushless (RBE Kollmorgen series) equipped with harmonic drive reduction of 100:1 (CSD series). Absolute encoders directly measure the motor angles.

2.1.2 Electronics and Sensors

The iCub robot is equipped with a rich set of sensors, either commercial or specifically realized for this platform, which enables the robot to exploit visual, proprioceptive, kinesthetic, vestibular, tactile and force sensing.

A gigabit Ethernet interface allows the main core (a PC104 computer) to communicate in an external network, typically used for intensive data processing. CAN-bus lines are employed for the communication between the boards and the PC104 (see Sect. 2.1.1.3). The PC104 board has the principal role of collecting and synchronizing all the sensory and motor data. To achieve this, a custom made board called CFW2 board is directly connected to the PC104 as the hardware interface between the overall 8 can-bus networks and the PC104. Motors are commanded with custom electronic boards mounting a Freescale 56F807 DSP. Two main control board have been specifically developed to control the motors: a BLL (BrushLessLogic unit) board, which control two brushless motors each board, and the MC4 (MotorControl4) board for the control of four DC motors for each board (see [4]). The control rate of these boards is 1 ms. Force sensors embed an home made electronic board for strain data acquisition (Strain board). The associated circuitry samples and amplifies up to 6 analog channels which can be used to measure the voltage across 6 strain gauges in a Wheatstone bridge configuration. The analog to digital converter (AD7685, 16 bit, 250 Ksps, SPI interface) is multiplexed (ADG658) on the 6 channels and amplified with a standard instrumentation amplifier (INA155). In our specific case, the sampling rate is 1 kHz. The Strain Board has a CAN-bus interface which allows its connection directly on the CAN-network. A more detailed overview of the low level hardware connection will be detailed shown in Chap. 6.

2.1.2.1 The Force/Torque Sensor

Force/torque sensors (FTSs) are used to measure the interaction of the robot with objects, human and the unstructured environment. The information measured by the FTSs is localized and represents the internal generalized forces acting at the FTS's

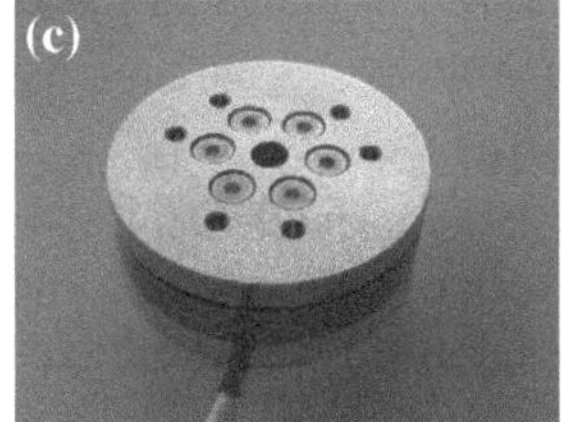

Fig. 2.8 The custom F/T sensor. **a** Picture of the sensing elements where the strain gages are placed. **b** The embedded board from which the measurements exits with sampled digital signal directly over a CAN-bus line, with a rate of 1 ms. **c** The assembled sensor

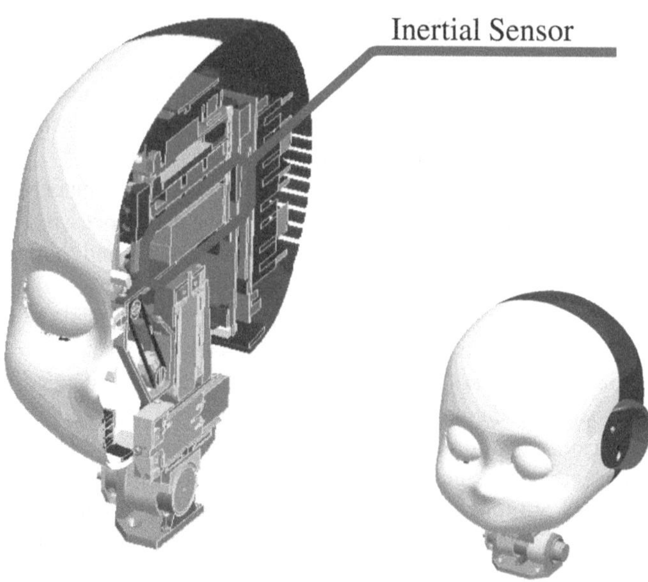

Fig. 2.9 3*d* section of the iCub head, which shows the *brain* of the robot, focusing on its vestibular system (the Inertial sensor [17])

location. This type of sensor is very important to study interaction tasks and their position along the kinematic chain of a robotic structure defines which information can be measured. All the four iCub limbs are equipped with custom made F/T sensors (see [16]). These sensors are placed in the upper part of the arm, between the shoulder and the elbow and in the legs, between the hip and the knee (see Fig. 2.8). These FTSs employ semiconductor strain gauges for measuring the deformation of the sensing elements. The signal conditioning and the analog to digital converters are embedded in the sensor. The data processing is performed on a 16 bit DSP from Microchip (dsPIC30F4013).

The solution adopted to place the FTSs in the iCub robot differs from the classical implementation which exploits FTSs placed at the end-effector level, typically adopted in industrial robots [14]. Specifically, the iCub F/T sensors are mounted proximally in each limb. This solution has different advantages:

- The F/T measurements give information about the arm internal dynamic.
- External forces applied on the arm (e.g. not only forces applied at the end-effector) can be sensed.
- Information about the actual joint torques can be extracted from the proximal F/T sensor.

Chapter 3 will show a formulation that allows to estimate internal wrenches, but also externally applied generalized forces given a set of distributed FTSs. Chapter 4 shows instead example of the application of the method, with particular attention to the iCub humanoid robot.

2.1.2.2 The Inertial Sensor

An Xsens MTx-28A33G25 [17] is placed on the head of the iCub robot, as shown in Fig. 2.9. The 3-DOF orientation tracker allows measuring the linear and angular acceleration and the angular velocity of the terminal link of the head. Chapter 3 describes a methodology to propagate the information along the links to perform the computation of kinematic quantities necessary to suddenly compute the dynamic of the system.

2.2 The iCub Software

An overview of the software architecture that allows commanding the robot is here introduced. A detailed description of the software and hardware architecture is also reported in Chap. 6, where some implementation issues related to the proposed methodology are also addressed.

2.2.1 Overview of the Hardware Components

A cluster of servers and standard PCs are interconnected through a 1 GB Ethernet and constitute the brain of the iCub robot. These machines are dedicated to run the part of software that is more computationally demanding (e.g. coordinated control, visual processing, learning). The low-level motor control is instead implemented on the DSPs embedded in the robot body.

2.2.2 The Yarp Framework

All the softwares which constitute the iCub high level *sensori-motor system* and cognitive architecture has been written using YARP [7]. YARP (Yet Another Robot Platform) is an open-source software framework that supports distributed computation under different operative systems (Windows, Linux and Mac OS) with the main goal of achieving efficient robot control. YARP facilitates code reuse and modularity by decoupling the programs from the specific hardware (using *Device Drivers*) and operative system (relying on the *OS wrapper* given by ACE [2, 11]) and by providing an intuitive and powerful way to handle inter-process communication (using *Ports* objects, which follows the Observer pattern [12]). YARP also provides mathematical (vectors and matrices) operations and image processing (basic *Image* class supporting IPL and OpenCV) libraries. The choice of this type of architecture finds its motivations in the fact that one single CPU, although powerful, can never be enough to cope with the demand of more and more computationally expensive applications. From here, the necessity of dividing the processes between more calculators connected together on the same local network. This distribution also encourage the development of modular and reusable code. Additional details about the software architecture are reported in Chap. 6.

2.2.3 *The iCub Interface*

The PC which is directly interfaced with the CAN-BUS lines executes a program (namely, *iCubInterface*) that manages the communication with the motor control boards, the various peripherals and sensors, and the user.

It allows retrieving the measurements of encoders, inertial sensor, FTSs and also of distributed tactile sensors over the entire body.

References

1. Cannata, G., Maggiali, M., Metta, G., & Sandini, G. (2008). An embedded artificial skin for humanoid robots. In *IEEE International Conference on Multisensor Fusion and Integration*, Seoul, Korea.
2. Gamma, E., Helm, R., Johnson, R., & Vlissides, J. M. (1994). *Design patterns: elements of reusable object-oriented software* (1st ed.). Reading, MA: Addison-Wesley Professional.
3. Kaneko, K., Kanehiro, F., Kajita, S., Hirukawa, H., Kawasaki, T., Hirata, M., et al. (2004). Humanoid robot hrp-2. In *Robotics and Automation, 2004. Proceedings. ICRA '04. 2004 IEEE International Conference on.*
4. Liralab.IT. (2010). The iCub user main page. http://eris.liralab.it
5. Maggiali, M., Cannata, G., Maiolino, P., Metta, G., Randazzo, M. & Sandini, G. (2008). Embedded distributed capacitive tactile sensor. In *Mechatronics 2008*. Limerick, Ireland.
6. Metta, G., Sandini, G., Vernon, D., Natale, L., & Nori, F. (2008). The iCub humanoid robot: an open platform for research in embodied cognition. In *PerMIS: Performance Metrics for Intelligent Systems Workshop*. Washington DC, USA.
7. Metta, G., Fitzpatrick, P., & Natale, L. (2006). Yarp: Yet another robot platform. *International Journal of Advanced Robotics Systems, special issue on Software Development and Integration in, Robotics*, *3*, 43–48.
8. Parmiggiani, A., Randazzo, M., Natale, L., Metta, G., & Sandini, G. (2009). Joint torque sensing for the upper-body of the iCub humanoid robot. In *International Conference on Humanoid Robots*, Paris, France.
9. RobotCub.org (2010). The RobotCub community portal. http://www.robotcub.org
10. Sandini, G., Metta, G., & Vernon, D. (2007). The iCub cognitive humanoid robot: An open-system research platform for enactive cognition. *50 Years of AI, LNAI 4850*, pp. (359–370). Berlin, Heidelberg: Springer-Verlag.

11. Schmidt, D. C. (2003). The adaptive communication environment. http://www.cs.wustl.edu/~schmidt/ACE.html
12. Schmidt, D. C. (2002). *C++ Network programming: mastering complexity using ACE and patterns*. Amsterdam: Addison-Wesley Longman.
13. Schmitz, A., Maggiali, M., Natale, L., Bonino, B., & Metta, G. (2010). A tactile sensor for the fingertips of the humanoid robot icub. Taipei, Taiwan, October 18–22, 2010.
14. Siciliano, B., & Villani, L. (2000). *Robot force control*. Norwell, MA, USA: Kluwer Academic Publishers.
15. Tsagarakis, N., Metta, G., Sandini, G., Vernon, D., Beira, R., Santos-Victor, J., et al. (2007). icub—the design and realization of an open humanoid platform for cognitive and neuroscience research. *International Journal of Advanced Robotics*, *21*(10), 1151–1175.
16. Tsagarakis, N., Becchi, F., Righetti, L., Ijspeert, A., & Caldwell, D. (2007). Lower body realization of the baby humanoid—iCub. In *IEEE/RSJ International Conference on Intelligent Robots and Systems*, San Diego, USA.
17. XsensMTx (2010). The MTx orientation tracker user manual. http://www.xsens.com/en/general/mtx
18. Tsagarakis, N., Metta, G., Sandini, G., Vernon, D., Beira, R., Santos-Victor, J., et al. (2007). icub - the design and realization of an open humanoid platform for cognitive and neuroscience research. *International Journal of Advanced Robotics*, *21*(10), 1151–1175.

Chapter 3
Propagation of Force Measurements Through MBSD

Abstract This chapter introduces the methodology proposed for building the dynamic model of single and multiple branches robotic mechanisms. It gives an overview of the formulation which allows to compute kinematic quantities and wrenches along the structure of generic serial mechanisms. In particular, the method allows to exploit measurements of sensors, to obtain a reliable estimation of internal generalized forces. The method can also be used to retrieve *virtual force/torque measurements* of externally applied generalized forces. The method is based on a graph formulation for the computation of the robot dynamics. This formulation allows both to obtain a very simple representation of the dynamic of multi-branched manipulators and to define *virtual measurements* of the internal dynamics by exploiting inertial and force/torque sensors.

3.1 Introduction to Rigid Body Dynamics

The kinematics and dynamics of robots can be described by set of elements, mutually constrained, that are characterized by properties of length, mass and inertia. These elements are typically called *rigid bodies*. The relative and absolute motion of a set of rigid bodies (or links) constituting a kinematic chain is described by its *joint constraints*. The type of mechanism constituting the joints define the constraint equations characterizing the relative motion between the links. Different detailed formulations of methods adopted for multi-body system dynamics can be found in [1]. Few of them are briefly introduced in Sects. 3.2.2 and 3.2.3.

M. Fumagalli, *Increasing Perceptual Skills of Robots Through Proximal Force/Torque Sensors*, Springer Theses, DOI: 10.1007/978-3-319-01122-6_3,

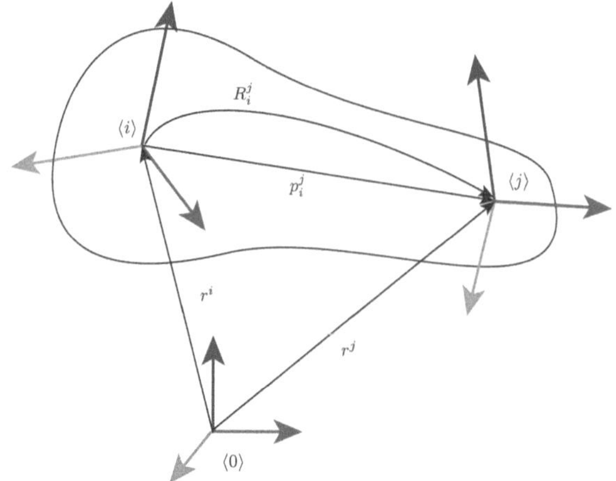

Fig. 3.1 Sketch of a rigid-body link

3.1.1 *Kinematics of the Rigid Body*

A rigid body can be defined as a set of points whose relative position does not change in time. It is in fact sufficient to describe the position and orientation of one of these points, with respect to a reference frame (generally indicated with $\langle \cdot \rangle$), that we can have the entire knowledge of the position and orientation of all the other points defining the rigid body. In other words, the position and orientation of a rigid body (see Fig. 3.1) is described by a point $p_i \in \mathbb{R}^3$ (actually the x, y and z coordinates with respect to a frame $\langle j \rangle$) and rotation matrix $R_i \in \mathbb{R}^{3\times3}$ with respect to the jth frame of reference (see [7] for more details). The position and orientation of a rigid body is therefore described by the homogeneous transformation between frame i and frame j. An homogeneous transformation is a matrix representation $\in \mathbb{R}^{4\times4}$, which defines the roto-translation matrix of one frame with respect to another, and takes the form:

$$T_i^j = \begin{bmatrix} a_{xx} & b_{xy} & c_{xz} & x \\ a_{yx} & b_{yy} & c_{yz} & y \\ a_{zx} & b_{zy} & c_{zz} & z \\ 0 & 0 & 0 & 1 \end{bmatrix} = \begin{bmatrix} R_i^j & p_i^j \\ 0 & 1 \end{bmatrix} \tag{3.1}$$

where $R_i^j \in \mathbb{R}^{3\times3}$ represent the rotation matrix of $\langle j \rangle$ with respect to $\langle i \rangle$, and $p_i^j \in \mathbb{R}^3$ is the distance vector from the origin of $\langle i \rangle$ to $\langle j \rangle$.

3.1.2 *Dynamics of the Rigid Body*

Let us consider a rigid body whose points P_i are characterized by a mass m_i ($i = 1, 2, \ldots, N$), with fixed relative distances between the points, and with center of mass (CoM) in a point $C = \left[C_x, C_y, C_z\right]^T$. Each point i constituting the rigid body is subject to the external force $f_{i,ext}$ and to the internal force $f_{i,int}$ exchanged with the other points defining the rigid body. The balance of forces acting on the rigid body is given by:

$$\sum_{i=1}^{N} m_i \frac{d^2 r_i}{dt^2} = M \frac{d^2 r_C}{dt^2} = \sum_{i=1}^{L} f_{i,int} + \sum_{j=1}^{H} f_{j,ext} = \sum_{i=1}^{H} f_{i,ext} = F_{ext} \tag{3.2}$$

The equation describes the linear motion of a rigid body. In particular it shows that the motion of a set of points of mass m_i, rigidly constrained together, is equal to the motion of one body of overall mass $M = \sum_i m_i$, subject to one external force $F_{ext} = \sum_k f_{k,ext}$.

Similarly, the angular momentum balancing follows:

$$M r_C^{\langle o \rangle} \frac{d^2 R_{\langle o \rangle}}{dt^2} + \frac{d(I\omega)}{dt^2} = \sum_{i=1}^{H} \tau_{\langle o \rangle, i} = \tau_{\langle o \rangle, ext} \tag{3.3}$$

being $\langle o \rangle$ a generic frame of reference with origin in $\langle o \rangle$, and I the inertia tensor of the rigid body.

3.2 Dynamics of Serial Mechanisms

The dynamic of a manipulator is described by a set of rigid bodies connected together by means of joints. The term *link* is used to refer to a rigid body. A link is characterized by a frame of reference defined on a point b moving with respect to another frame a and whose relative position can change according to the definition of the joint constraint. A serial manipulator is a sequence of links connecting the base link to terminal link, where each link is connected to the other by means of actuated joints.

In this manuscript we refer to the sole revolute joints. A revolute joint is a set of five scalar equations which defines a constraint between link a and link b by defining the kinematic relationship between point p_a (rigidly constrained to $\langle a \rangle$) where the joint is located, and p_b (rigidly constrained to $\langle b \rangle$). With reference to figure Fig. 3.2, a revolute joint introduces a constraint that forces the position p_a of the origin of frame $\langle a \rangle$ and the position p_b of the origin of frame $\langle b \rangle$ to have a constant distance vector r_a^b (which correspond to three scalar equation $\Psi^p(p_a, p_b) = r_a^b$). Moreover, it introduces constraints that force the orientation of the axis z_a of frame $\langle a \rangle$ and z_b of frame $\langle b \rangle$ to have the same direction (which correspond to two scalar equation $\Psi^z(z_a, z_b) = 0$). The overall number of relative degrees of mobility of the kinematic couple of the two links is therefore one, and refers to the relative angular position of the axis x_a of frame $\langle a \rangle$ and x_b of frame $\langle b \rangle$, that we call θ.

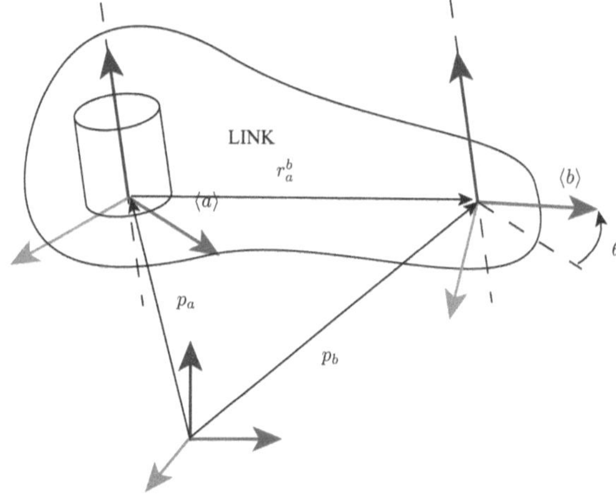

Fig. 3.2 Sketch of a link with revolute joint

Note that different joints introduce different constraint equation. For the definition of other kinematic couples, the interested reader should refer to [1].

3.2.1 Notation

Let us introduce the notation of symbols that will be used to denote kinematic and dynamic quantities. With reference to Fig. 3.3 we call:

- $\langle \cdot \rangle$ generic Cartesian reference frame, where $\langle i \rangle$ refers to the reference frame attached to the ith joint
- θ_i the angle associated to the ith joint. The vector of joint coordinates of the manipulator is denoted $\theta \in \mathbb{R}^n$
- $\ddot{p}_i$ $\in \mathbb{R}^3$, denotes the linear acceleration of frame $\langle i \rangle$h
- $\omega_i, \dot{\omega}_i$ $\in \mathbb{R}^3$, the angular velocity and acceleration of $\langle i \rangle$

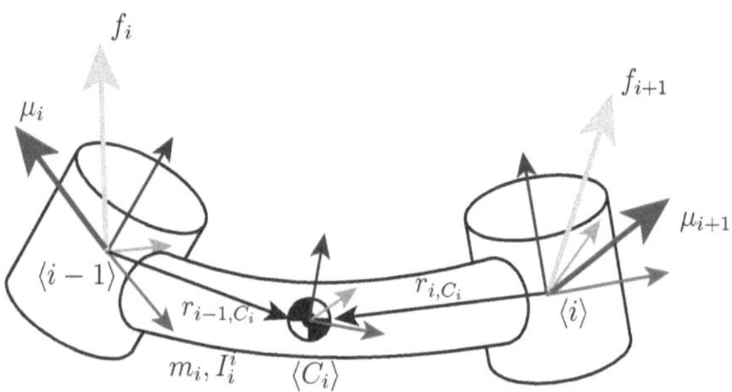

Fig. 3.3 Notation for the ith link of a kinematic chain

v_a given $v \in \mathbb{R}^n$ a generic n-dimensional vector, v_a is v expressed in $\langle a \rangle$
R_a^b the $SO(3)$ rotation matrix from $\langle a \rangle$ to $\langle b \rangle$
$r_{a,b}$ distance vector r from $\langle a \rangle$ to $\langle b \rangle$
C_i $\in \mathbb{R}^3$ the coordinate vector of the center of mass of link ith, with respect to $\langle i \rangle$
z_i z-axis of $\langle i \rangle$, aligned with the axis of rotation of joint i
m_i mass associated with the ith link
$\bar{I}_i^i$ $\in \mathbb{R}^{3\times 3}$, defined with respect to the center of mass oriented as the frame $\langle i \rangle$, represent the inertia tensor of the ith link
f_i $\in \mathbb{R}^3$, represent the forces applied on $\langle i \rangle$, that link $i+1$ exerts on the ith link
μ_i $\in \mathbb{R}^3$, represent the moment applied on $\langle i \rangle$, that link $i+1$ exerts on the ith link
τ_i $\in \mathbb{R}$ the joint torque, i.e. the component of μ_i along z_i
w_i $\in \mathbb{R}^3$ the wrench (or generalized forces) $w = \begin{pmatrix} f \\ \mu \end{pmatrix}$ applied on $\langle i \rangle$, that link $i+1$ exert on link i.

The formulation of the kinematics and the dynamics of multi-body systems which is presented in next sections will refer to his notation. Different formalisms allow to describe the kinematics and dynamics of multi-body systems. Section 3.2.2 shows some of the methods to describe the kinematics of manipulators, while Sect. 3.2.3 gives an idea on different formulations to describe the robot dynamics. Based on the notation above, a general formulation which addresses the kinematics and dynamics computation for multiple branched robotic system, adopting graph theory, will be presented in Sect. 3.3.

3.2.2 *Kinematic Description: The Denavit-Hartenberg Notation*

The kinematic formulation that will be adopted here to describe the kinematics of manipulators is the *Denavit-Hartemberg* (DH) notation (see [7]). Compared to other methods (see [1]), the DH notation has the advantage to be procedural and easy. It allows the description of the kinematics of links by means of four parameter which define the relative position and orientation of sequential reference frames. The DH method is also suitable for the representation of the kinematics of multi-branched robotic mechanisms.

Among the disadvantages it should be noted that the DH representation introduces constraints on the choice of reference frames. The introduction of constraints in the choice of the reference frames lead to unique definition of particular links such as the base link and the terminal ones.

When revolute joints are considered, the method consists in defining one frame for each link, according to some rules (see [7]) which allow to represent the links in the form of an homogeneous transformation:

$$T_i^{i-1} = \begin{bmatrix} cos(\theta_i) & -sin(\theta_i)cos(\alpha_i) & sin(\theta_i)sin(\alpha_i) & a_i cos(\theta_i) \\ sin(\theta_i) & cos(\theta_i)cos(\alpha_i) & -cos(\theta_i)sin(\alpha_i) & a_i sin(\theta_i) \\ 0 & sin(\alpha_i) & cos(\alpha_i) & d_i \\ 0 & 0 & 0 & 1 \end{bmatrix} \quad (3.4)$$

being θ_i the relative rotation around the joint axis z_i of frame i with respect to frame $i-1$. α_i, a_i and d_i are fixed parameters describing the link kinematic.

3.2.3 Dynamic Description: The RNEA

The dynamic description of a mechanical system can be achieved adopting different methods. Some relies on energetic formulation and conservation theories, such as the *Lagrange formulation*, while others relies on the balancing of the forces and moments acting on the rigid bodies, as the *Recursive Newton-Euler Algorithm* (RNEA). In the first case, the formulation gives origin to sets of equations in closed form that can be mainly employed for the analysis of control scheme and system properties. The latter instead are mostly used for implementation, due to the intrinsically recursive formulation, which allows fast numerical computation of the inverse dynamic, which is essential for real-time control systems. In this section, the analysis focuses on the RNE algorithm. We show the formulation of the classical method, and its adaptation to graph formulation.

The analysis of systems dynamic typically consists in two main steps: first, kinematic quantities, actually the linear position and orientation of frames, and corresponding velocities and accelerations of the links must be determined; secondly the computation of dynamic quantities is performed.

The classical Newton-Euler algorithm performs these computation through a recursion in two directions along the kinematic chain. A forward recursion allows to determine the kinematic quantities referred to the links' frame of reference, actually their position, velocities and acceleration. A backward recursion instead performs the computation of wrenches acting on the frame of references of the links. In the backward recursion, forces and moments acting on each link are determined through equations which consider the balancing of these quantities on the rigid body (as already introduced in Sect. 3.1.2).

The classical RNE equation follows here: adopting the Denavit-Hartenberg notation (once again refer to [8] for details), define a set of reference frames $\langle 0 \rangle$, $\langle 1 \rangle$, ..., $\langle n \rangle$ attached at each link. Considering a grounded manipulator, set the velocities and acceleration of the base frame as: $\ddot{p}_0 = -\mathbf{g}$, $\omega_0 = [0, 0, 0]$ and $\dot{\omega}_0 = [0, 0, 0]$. The Newton-Euler kinematic step consists in the propagation of velocity/acceleration information from the base to the end-effector (forward kinematics), considering the relative velocities and acceleration between subsequent links, induced by joint motion:

$$
\begin{aligned}
\omega_{i+1} &= \omega_i + \dot{\theta}_{i+1} z_{i+1}, \\
\dot{\omega}_{i+1} &= \dot{\omega}_i + \ddot{\theta}_{i+1} z_{i+1} + \dot{\theta}_{i+1} \omega_i \times z_{i+1}, \\
\ddot{p}_{i+1} &= \ddot{p}_i + \dot{\omega}_i \times r_{i,i+1} + \omega_{i+1} \times (\omega_{i+1} \times r_{i,i+1}),
\end{aligned}
\tag{3.5}
$$

where z_{i+1} represent the z-axis of frame $i+1$. Measuring $\theta_i, \dot{\theta}_i, \ddot{\theta}_i$ the above equations can be iterated to retrieve the i-th link angular velocity and acceleration (ω_i, $\dot{\omega}_i$) and linear acceleration ($\ddot{p}_i$).

Considering that the system is moving freely (i.e. without interacting with the environment), the robot dynamics is computed starting from the end-effector (where f_{i+1} and μ_{i+1} are set equal to zero) to the base. For each link, the force and torque components on joints which allow the maintenance of the system equilibrium are the computed as:

$$
\begin{aligned}
f_i &= f_{i+1} + m_i \ddot{p}_{C_i}, \\
\mu_i &= \mu_{i+1} - f_i \times r_{i-1,C_i} \\
&\quad + f_{i+1} \times r_{i,C_i} + \bar{I}_i^i \dot{\omega}_i + \omega_i \times (\bar{I}_i^i \omega_i),
\end{aligned}
\tag{3.6}
$$

where

$$
\ddot{p}_{C_i}^i = \ddot{p}_i^i + \dot{\omega}_i^i \times r_{i,C_i}^i + \omega_i^i \times (\omega_i^i \times r_{i,C_i}^i).
\tag{3.7}
$$

Assuming the system dynamical parameters are known (m_i, $\bar{I}_i^i$, r_{i-1,C_i}, r_{i,C_i}), wrenches are thus propagated to the base frame of the manipulator so as to retrieve f_0 and μ_0.

3.3 Dynamics of Multiple Branched Mechanisms: A Graph Formulation

Graph theory has been extensively used to represent mechanical systems (see [3, 9]) and kinematic chains, producing compact and clear models, in matrix forms with beneficial properties (e.g. branch-induced sparsity, shown in [2]) when the connectivity among its elements is expressed. There is not a unique choice for a graph representing a chain: for example, in [4] graphs are undirected, nodes and arcs represent bodies and joints respectively; the resulting graph is undirected (i.e. non-oriented), but nodes are "labeled" according to a "regular numbering scheme".

This section presents the theoretical framework of the *Enhanced Oriented Graphs* (EOG), applied to the computation of both internal and external wrenches applied to single and multiple branches, generally non-grounded, kinematic chains. The proposed method is independent on the equation exploited for performing the calculation. Within this manuscript the *recursive Newton-Euler algorithm* [3, 8] will be adopted, but it must be noted that this choice is not the unique for computing the inverse dynamic by means of the graph formulation that is proposed in this manuscript. In the methodology that will be presented [5, 6], kinematic chains are represented as graphs. Differently to classical approach for inverse dynamic computation exploiting graph theories, the graph is enhanced with specific nodes representing both known and unknown (kinematic or dynamic) variables. Therefore, nodes representing the known variables (i.e. sensors placed along the kinematic tree) and nodes representing the unknown variables (i.e. virtual sensors giving an estimate of the unknown external wrenches acting on a certain link) will be added to the formulation. Remarkably, not all the unknowns will be specified a-priori (e.g. contacts at arbitrary locations might appear and other contacts might be removed) and therefore the graph structure will be adapted accordingly.

The computations of the system dynamics are performed through a pre-order and a post-order traversal visit of the graph itself, depending on the quantity to be computed. Pre-order traversal visit of a graph means that the propagation of the information starts from the root towards all the leaves of the tree. Post-order traversal instead means that the propagation of the information starts from the leaves and sub-branches, to finally visit the root. Figure 3.4 describe these two possibilities for visiting a graph. The resulting graph allows to dynamically represent the unknown quantities (both kinematic and dynamic) of a link.

The description of the chain infact evolves on the basis of the position of the unknown variables. This reflects in the fact that the way the graph is visited during the recursion modifies the direction along which the information is propagated in the graph. In order to cope with this evolving representation another difference with respect to previous graphical representations is introduced. The kinematic chain will be represented as an *oriented* graph: the direction along which edges are traversed will determine the recursion formula to be employed.

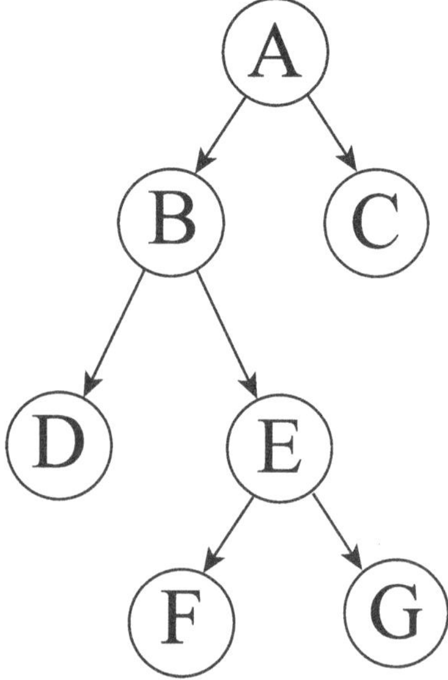

Fig. 3.4 Example of multiple branched graph structure. When the graph is visited in pre-order, from the root A the visited nodes are B, D, E, F, G, C. When post-order, the visiting order is F, G, E, D, B, C and A

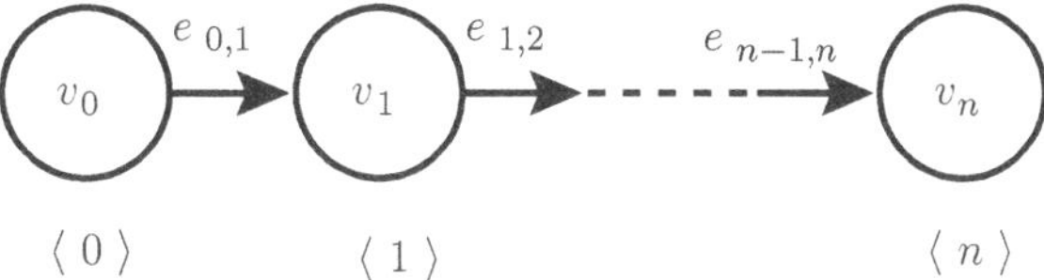

Fig. 3.5 An open chain represented as a graph

3.3.1 The Enhanced Graph Representation

Let us consider an open (single or multiple branches) kinematic chain with n DOF composed of $n + 1$ links (see Fig. 3.5). The ith link of the chain is represented by a vertex v_i (sometimes called node). A hinge joint between the link i and the link j (i.e. a rotational joint) is represented by an oriented edge $e_{i,j}$ connecting v_i with v_j (see Fig. 3.6). The orientation of the edge can be either chosen arbitrarily (it will be clear later on that the orientation simply induces a convention) or it can follow from the exploration of the kinematic tree according to the "regular numbering scheme" [3], which induces a parent/child relationship such that each node has a unique input edge and multiple output edges. As a convention, it is here assumed that each joint has an associated reference frame with the z-axis aligned with the rotation axis; this frame will be denoted $\langle e_{i,j} \rangle$. In kinematics, an edge $e_{i,j}$ from v_i to v_j represents the fact that $\langle e_{i,j} \rangle$ is fixed in the ith link. In dynamics, $e_{i,j}$ represents the fact that the dynamic equations will compute (and make use of) $w_{i,j}$, i.e. the wrench that the ith link exerts on the jth link, and not the equal and opposite reaction $-w_{i,j}$, i.e. the wrench that the jth link exerts on the ith link (further details in Sect. 3.4). In order to simplify the computations of the inverse dynamics on the graph (as will be shown in Sect. 3.4), kinematic and dynamic measurements have been explicitly represented through additional types of nodes (see Fig. 3.6). In particular, the graph representation has been enhanced with a new set of graphical symbols: a triangle to represent kinematic quantities (i.e. velocities and acceleration of links), and a rhombus for wrenches (i.e. force sensors measurements within a link), as shown in Fig. 3.6. Moreover these symbols have been further divided into *known* quantities to represent sensors measurements, and *unknown* to indicate the quantities to be computed.

3.3.2 Kinematics

Kinematic variables can in general be measured by means of gyroscopes, accelerometers, or simply inertial sensors. When attached on link ith, these sensors provide angular and linear velocities and accelerations (ω, $\dot{\omega}$, $\dot{p}$ and $\ddot{p}$) at the specific location where the sensor is located. These measurement are represented in the graph with a *black triangle* (▼) and an additional edge from the proper link where the sensor is attached to the triangle. Note that, according to the kinematic convention previously

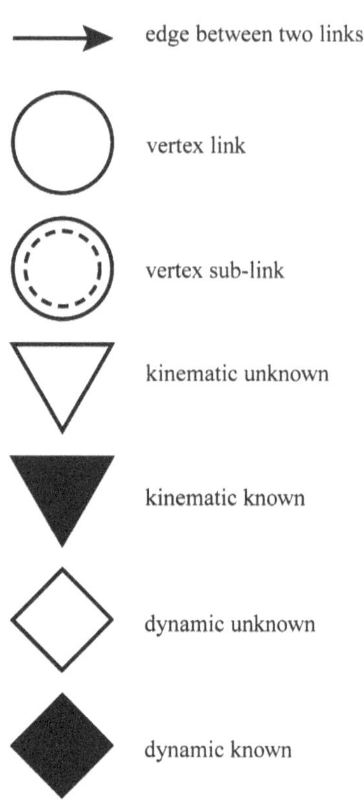

Fig. 3.6 The notation introduced to represent nodes, sub-nodes, edges, known and unknown kinematic and dynamic variables in graphs

mentioned, an edge $e_{i,j}$ is fixed on the ith link. Therefore a sensor fixed in the ith link, will be represented by $e_{i,s}$, i.e. an edge from the link to the sensor (see Fig. 3.7a, b). Also in this case, the reference frame associated to the edge corresponds to the reference frame of the sensor. Similarly, an unknown kinematic variable is represented with a *white triangle* ($\triangledown$) with an associated edge going from the link (where the unknown kinematic variable is attached) to the triangle. The reference frame associated to the edge will determine the characteristics of the retrieved unknown kinematic variables as it will be clear in Sect. 3.4.

3.3.3 Dynamics

Similarly, we introduce two new types of nodes with a rhomboidal shape (see Fig. 3.6): *black rhombi* (♦) to represent known (i.e. measured) wrenches, *white rhombi* (◊) to represent unknown wrenches which need to be computed. The reference frame associated to the edge will be the location of the applied or unknown wrench (see Fig. 3.8). Remarkably, there is not a fixed rule to determine the

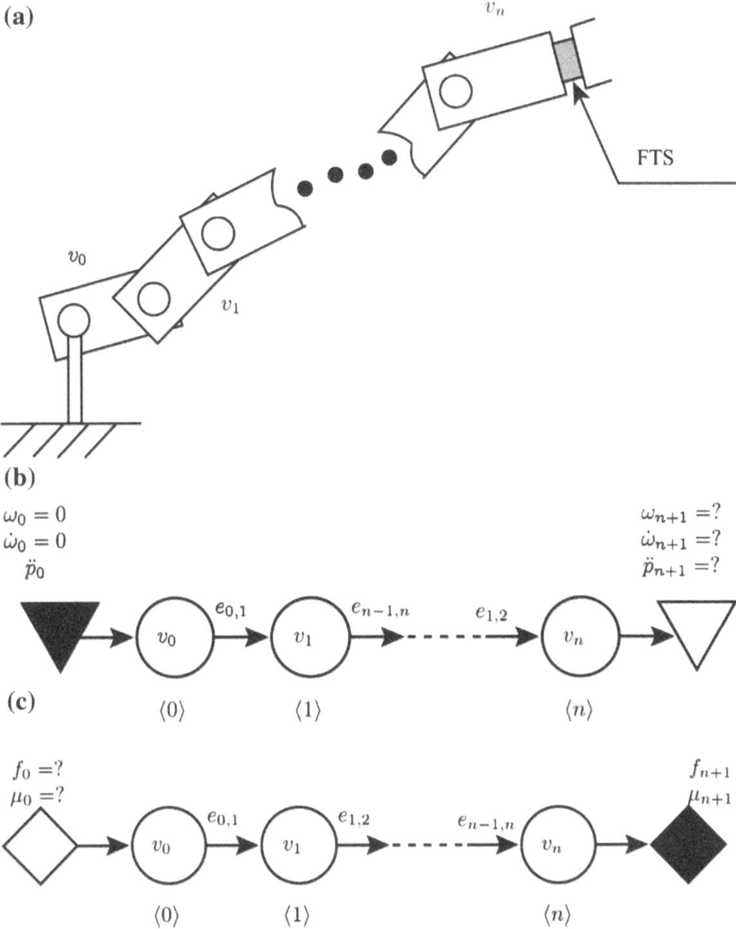

Fig. 3.7 **a**: sketch of a serial robotic structure with a FTS placed at the end-effector; kinematic quantities are known at the base of the robot, where $\omega_0 = \dot{\omega}_0 = 0$ and $\ddot{p}_0 = \mathbf{g}$. **b**: graph representing the kinematic EOG of **a**. **c**: graph representing the dynamic EOG of **a**

orientation of the edge connecting the rhombi to the graph: according to this convention for representing the wrenches, the edge can be either directed from the rhombus to the link or *vice versa* depending on the variable we are interested in representing (i.e. the wrench from the link to the external environment or the equal and opposite wrench from the environment to the link). It is important to point out that, whereas the position of ⧫ is static within the graph (because sensors are fixed in the manipulator), the location of ◊ instead can be dynamic (contact point locations are dynamically detected by the distributed tactile sensor). If a contact moves along a chain, the graph is accordingly modified. This rule shows a big benefit of the EOG, which dynamically adapts in response to the location of the unknown external wrenches. Within this

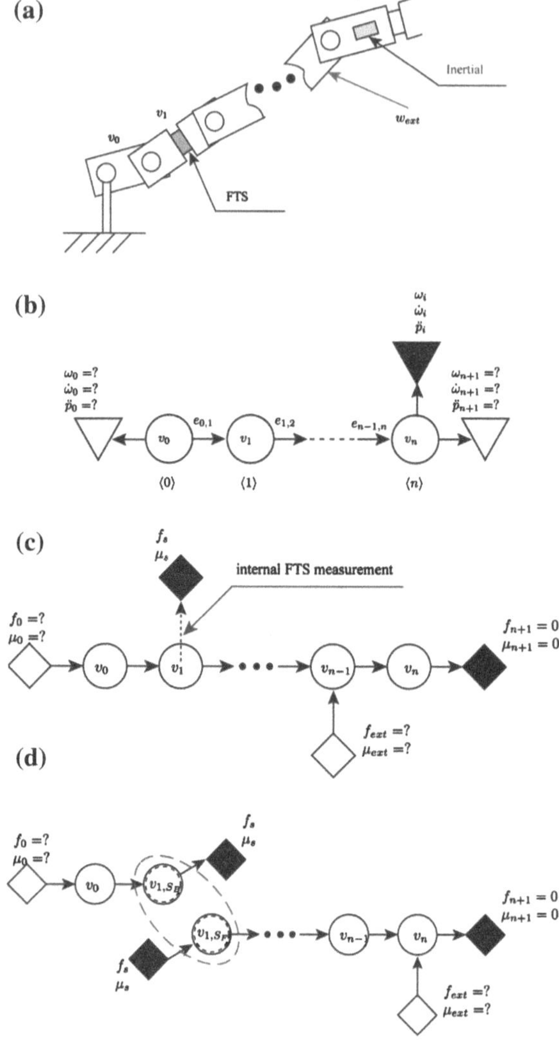

Fig. 3.8 **a**: sketch of a serial robotic structure with a FTS placed on a proximal link; an inertial sensor is placed on the nth link. **b**: graph representing the kinematic EOG of **a**. **c**: graph representing the dynamic EOG of **a**, before the division into two subchains. **d**: graph representing the two dynamic EOGs of **a**, consequence of the division of the link mounting the FTS into two sublinks; In this case, two unknown wrenches can be determined, one for each subchain

representation, embedded FTS can be inserted by "cutting" the manipulator chain where the FTS is located and creating two virtual "sub-links" from the link physically hosting the sensor (see also Fig. 3.9). The EOG is then split into two sub-graphs, where black rhombi (♦, i.e. known wrenches representing the FTS measures, one per graph) are introduced and attached to the sub-links. In practice, suppose that an FTS is placed in the i_Sth link (see Fig. 3.8a). Let $\langle s \rangle$ be the frame associated to the sensor. The sensor virtually divides link i_S into two "sub-links" (hereafter denoted forward v_{i,S_F} and the backward v_{i,S_B} sub-links, as shown in Fig. 3.8d). The sensor therefore measures the wrench exchanged between the "forward" and the "back-

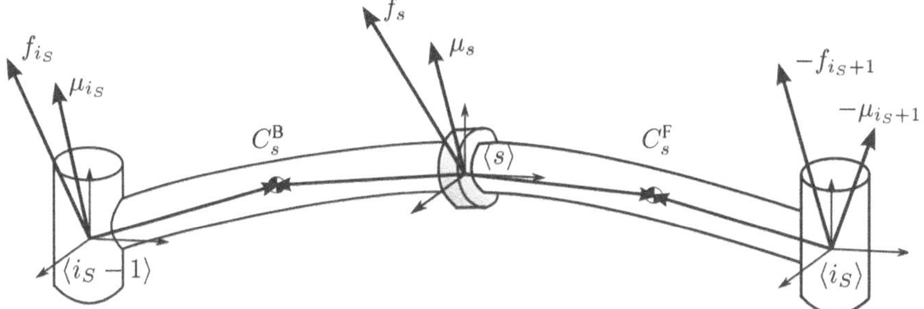

Fig. 3.9 A representation of an FTS within the i_Sth link. Note that the sensor divides the link into two sub-links, each with its own dynamical properties. In particular, it is evident that the center of mass (COM) of the original link, C_{i_S}, differs from C_s^{F}, C_s^{B}, i.e. the COM of the two "sub-links"

ward" sub-links (this will be represented by two rhomboidal nodes). Under these considerations, the FTS within a link is represented by splitting the node associated to the link into two sub-nodes (with suitable dynamical properties, see Fig. 3.8d). Two known wrenches in the form of black rhombi are then attached to the sub-nodes, with suitable edges whose associated reference frame is $\langle s \rangle$ for both edges.

3.4 Exploiting the RNEA for EOG

The graphical representation proposed in Sect. 3.3 can be adopted to represent the flow of information within kinematic chains (see Fig. 3.10), which are necessary to perform the computation of the internal dynamics of a kinematic chain provided with sufficient tactile, proprioceptive, haptic and inertial sensors. In particular, in this section we describe how to compute both kinematic and dynamic variables, associated to the edges of the graphical representation, for the general case of (floating) multiple branches kinematic chains. The method shown hereafter adopts the Newton-Euler formulas, but any other recursive formulation might be employed to perform the computation.

A first recursion on the graph (pre-order traversal) will compute the linear acceleration ($\ddot{p}$) and the angular velocity and acceleration (ω, $\dot{\omega}$) for each of the reference frames associated to the edges of the graph. This procedure practically propagates the information coming from a single inertial sensor to the entire kinematic chain. At each step, the values of ($\ddot{p}$, ω, $\dot{\omega}$) for a given link are propagated to neighbor links by exploiting the encoder measurements and a kinematic model of the chain. A second recursion (post-order traversal) will compute all the (internal and external) wrenches acting on the chain at the reference frames associated with all the edges in the graph. In this case, Newton-Euler equations are exploited to propagate force information along the chain. At each step, all but one wrench acting on a link are

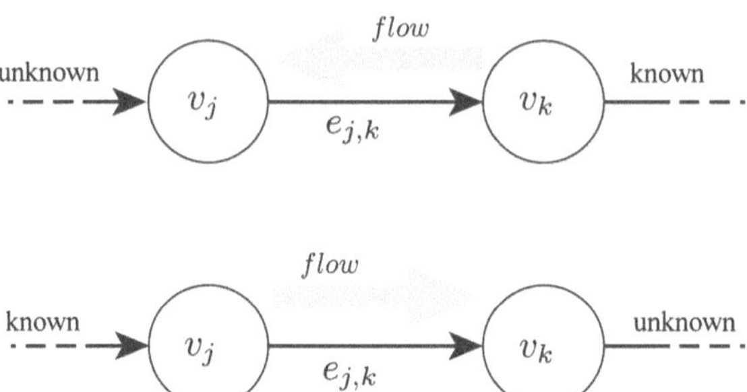

Fig. 3.10 The basic operation for propagating information across an EOG. Given v_j we assume to know ω_j, $\dot{\omega}_j$, $\ddot{p}_j$. This information can then be propagated to all the connected nodes. If v_k is connected to v_j by $e_{j,k}$ (i.e. the edge is directed from v_j to v_k) then we can compute ω_k, $\dot{\omega}_k$, $\ddot{p}_k$ using (3.8) (just replace $i+1$ with k). If v_k is connected to v_j by $e_{k,j}$ (i.e. the edge is directed from v_k to v_j) then we can compute ω_k, $\dot{\omega}_k$, $\ddot{p}_k$ using (3.8) (just replace $i-1$ with k). Similar considerations can be done for dynamic variables

assumed to be known and the remaining unknown wrench is computed exploiting a dynamic model of the link and the output from the kinematic recursion.

When computing wrenches, the computation of one unknown can only be performed for each subchain. Practically speaking, each chain represents one 6-dimentional equation, where at least one 6-dimentional unknown can be determined. If more than one unknown is present on a subchain, an infinite of solution can be found, and thus the method cannot be applied without introducing some assumptions on the contacts.

3.4.1 Kinematics

We here describe the basic equations for propagating the kinematic information within the graph. The proposed formulation is capable of exploiting multiple, dynamically inserted, inertia sensors to propagate the kinematic information from the sensors to the surrounding links. Therefore the flow of kinematics cannot be predefined but needs to be dynamically adapted to the current structure of the EOG. The basic step here described consists in propagating the kinematic information associated to an edge connected to a node v to all the other edges connected to v. As usual, for each edge i we consider the associated reference frame $\langle i \rangle$. Referring to Fig. 3.11a–c we assume that knowing the linear acceleration ($\ddot{p}_j$) and the angular velocity and acceleration ($\omega_j, \dot{\omega}_j$) of the reference frame $\langle j \rangle$ we want to compute the same quantities for the frame $\langle i \rangle$ sharing with $\langle j \rangle$ a common node v. Figure 3.11a represents the case where the edge i exits v but the edge j enters v; recalling the kinematic meaning of the edge directions, the sketch in Fig. 3.11a represents a situation where $\langle i \rangle$ is attached to v while $\langle j \rangle$ is rotated by the joint angle θ_j around z_j. The situation is exactly the one we have in the classical Denavit-Hartenberg forward kinematic description and therefore we have[1] (see [8]):

$$\begin{aligned}\omega_i &= \omega_j + \dot{\theta}_j z_j,\\ \dot{\omega}_i &= \dot{\omega}_j + \ddot{\theta}_j z_j + \dot{\theta}_j \omega_j \times z_j,\\ \ddot{p}_i &= \ddot{p}_j + \dot{\omega}_i \times r_{j,i} + \omega_i \times (\omega_i \times r_{j,i}),\end{aligned} \tag{3.8}$$

where z_j and θ_j indicate the rotational axis and the angular position of the joint associated to the edge j. Similarly, Fig. 3.11b represents the case where the edge i enters v but the edge j exits the node; therefore Fig. 3.11b represents a situation where $\langle j \rangle$ is attached to v while $\langle i \rangle$ is rotated by the joint angle θ_i. The situation is exactly the opposite encountered in classical Denavit-Hartenberg so that we have:

$$\begin{aligned}\omega_i &= \omega_j - \dot{\theta}_i z_i,\\ \dot{\omega}_i &= \dot{\omega}_j - \ddot{\theta}_i z_i - \dot{\theta}_i \omega_j \times z_i,\\ \ddot{p}_i &= \ddot{p}_j - \dot{\omega}_j \times r_{i,j} - \omega_j \times (\omega_j \times r_{i,j}).\end{aligned} \tag{3.9}$$

Finally, Fig. 3.11c represents the case where both $\langle i \rangle$ and $\langle j \rangle$ are attached to the link represented by v. In this case, continuity formulas are obtained putting $\dot{\theta}_i = 0$ and $\ddot{\theta}_i = 0$ in Eq. 3.8 (or equivalently Eq. 3.9):

$$\begin{aligned}\omega_i &= \omega_j,\\ \dot{\omega}_i &= \dot{\omega}_j,\\ \ddot{p}_i &= \ddot{p}_j + \dot{\omega}_i \times r_{j,i} + \omega_i \times (\omega_i \times r_{j,i}).\end{aligned} \tag{3.10}$$

These rules can be used to propagate kinematic information across different edges connected to the same node. The only situation which cannot be solved is the one where all edges enter the node v, i.e. none of the associated reference frames is fixed to the link v. We can handle these cases *a posteriori* by defining a new arbitrary reference frame $\langle v \rangle$ attached to the link. In our formalism, this is achieved by adding a kinematic unknown (∇) and an edge from v to ∇ with associated frame $\langle v \rangle$. It is remarkable here that, if the edge directions are chosen according to a "regular numbering scheme" as proposed in Sect. 3.3.1, each edge will have a unique ingoing edge and multiple outgoing edges. The only nodes with no outgoing edges will be the ones corresponding to the leaves of the kinematic tree (typically the end-effectors). For these nodes, we will add a kinematic unknown (∇) and an edge from v to ∇ with associated frame $\langle v \rangle$ (typically the end-effector reference frame of the classical Denavit-Hartenberg notation).

[1] In the classical recursive kinematic computation (as in [8]) there is a one-to-one correspondence between links and joints (see Fig. 3.3) thus resulting in a kinematic equations slightly different from Eq. 3.8. Classically, the ith link has two joints and associated reference frames $\langle i \rangle$ and $\langle i-1 \rangle$, respectively. Only $\langle i \rangle$ is attached to the ith link while $\langle i-1 \rangle$ is attached to the link $i-1$. The rotation between these two links is around the z-axis of $\langle i-1 \rangle$ by an angle which is denoted θ_i and therefore the analogous of Eq. 3.8 in [8] refer to $\dot{\theta}_i$ in place of $\dot{\theta}_j$ and z_{i-1} in place of z_i. In our notation, we get rid of this common labeling for joints and links by explicitly distinguishing the link represented with the node v and the attached joints represented with the edges $i, j, \ldots$ and associated frames $\langle i \rangle, \langle j \rangle, \ldots$ whose axes are therefore $z_i, z_j, \ldots$ with associated angles θ_i, θ_j.

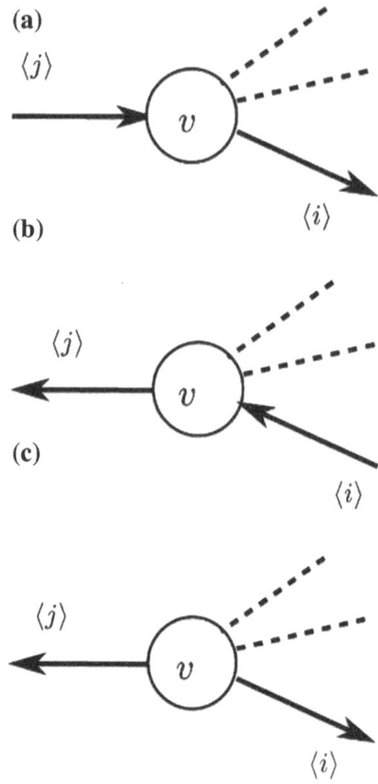

Fig. 3.11 The three cases accounting for the exchange of kinematic information

3.4.2 Dynamics

We here describe the basic equations for propagating the dynamic information within the graph. Also in this case, the flow of dynamical information cannot be predefined because the graph structure continuously changes according to the position of the applied external wrenches (as detected by the distributed tactile sensor). The basic step proposed in this section assumes that all but one wrench acting on a link are known and the remaining unknown wrench is computed by using the Newton-Euler equations. Using the graph representation, a node v with all its edges represents a link with all its joints. As proposed in Sect. 3.3.1, at each edge $e_{u,v}$, we can associate the wrench $w_{e_{u,v}}$ that u exerts on v. At each edge $e_{v,u}$ we can associate the wrench $w_{e_{v,u}}$ that v exerts on u. The Newton-Euler equations for the link v can therefore be written as follows:

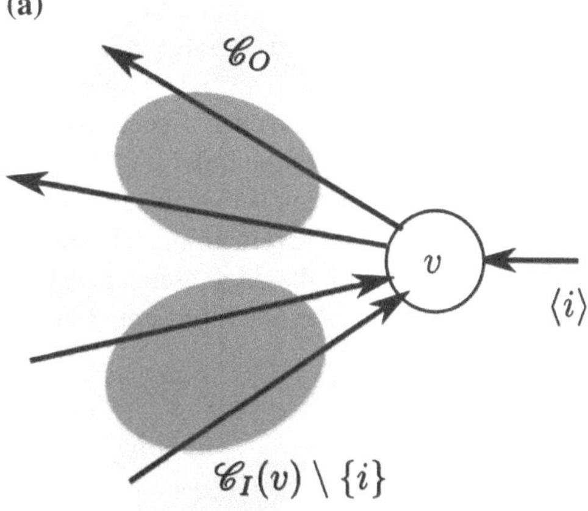

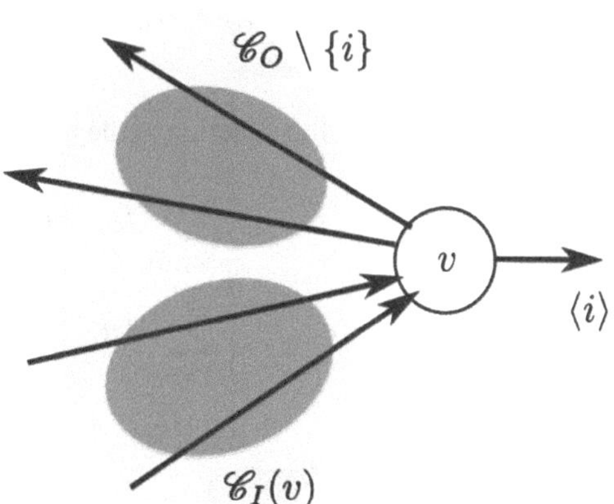

Fig. 3.12 The two cases accounting for the exchange of dynamic information

$$\begin{aligned} \sum_{e_I \in \mathscr{C}_I(v)} f_{e_I} - \sum_{e_O \in \mathscr{C}_O(v)} f_{e_O} &= m_v \ddot{p}_{C_v}, \\ \sum_{e_I \in \mathscr{C}_I(v)} \left(\mu_{e_I} + f_{e_I} \times r_{e_I, C_v}\right) & \\ - \sum_{e_O \in \mathscr{C}_O(v)} \left(\mu_{e_O} + f_{e_O} \times r_{e_O, C_v}\right) &= \bar{I}_i \dot{\omega}_i + \omega_i \times (\bar{I}_i \omega_i), \end{aligned} \tag{3.11}$$

where[2]

$$\ddot{p}_{C_v} = \ddot{p}_i + \dot{\omega}_i \times r_{i,C_v} + \omega_i \times (\omega_i \times r_{i,C_v}), \tag{3.12}$$

[2] With slight abuse of notation, the vector connecting the generic frame $\langle \star \rangle$ to the one placed on the center of mass C_v of the vth link is indicated with $r_{\star, C_v}$.

and where $\mathscr{C}_I(v)$ is the set of ingoing edges, $\mathscr{C}_O(v)$ is the set of outgoing edges and where the index i refers to any edge in $\mathscr{C}_O(v)$ (necessarily non-empty in consideration of what we discussed in Sect. 3.4.1). In other terms, recalling the kinematic meaning of outgoing edges, i is an edge associated with any of the arbitrary reference frames $\langle i \rangle$ fixed with respect to the link v. As anticipated, Eq. 3.11 can be used to propagate the dynamic information across the graph. Assuming that all but one wrench acting on a link are known, the remaining unknown wrench can be computed with Eq. 3.11. Let us denote with i the edge associated with the unknown wrench. If $i \in \mathscr{C}_I(v)$, then the situation is the one represented in Fig. 3.12a and we have:

$$\begin{aligned} f_i &= -\sum_{\substack{e_I \in \mathscr{C}_I(v) \\ e_I \neq i}} f_{e_I} + \sum_{e_O \in \mathscr{C}_O(v)} f_{e_O} + m_v \ddot{p}_{C_v}, \\ \mu_i &= -f_i \times r_{i,C_v} - \sum_{\substack{e_I \in \mathscr{C}_I(v) \\ e_I \neq i}} \left(\mu_{e_I} + f_{e_I} \times r_{e_I,C_v}\right) \\ &\quad + \sum_{e_O \in \mathscr{C}_O(v)} \left(\mu_{e_O} + f_{e_O} \times r_{e_O,C_v}\right) + \bar{I}_i \dot{\omega}_i + \omega_i \times (\bar{I}_i \omega_i). \end{aligned} \tag{3.13}$$

If $i \in \mathscr{C}_O(v)$, then the situation is the one represented in Fig. 3.12b and we have:

$$\begin{aligned} f_i &= \sum_{e_I \in \mathscr{C}_I(v)} f_{e_I} - \sum_{\substack{e_O \in \mathscr{C}_O(v) \\ e_O \neq i}} f_{e_O} - m_v \ddot{p}_{C_v}, \\ \mu_i &= -f_i \times r_{i,C_v} + \sum_{e_I \in \mathscr{C}_I(v)} \left(\mu_{e_I} + f_{e_I} \times r_{e_I,C_v}\right) \\ &\quad - \sum_{\substack{e_O \in \mathscr{C}_O(v) \\ e_O \neq i}} \left(\mu_{e_O} + f_{e_O} \times r_{e_O,C_v}\right) - \bar{I}_i \dot{\omega}_i - \omega_i \times (\bar{I}_i \omega_i). \end{aligned} \tag{3.14}$$

Note that, with reference to Eqs. 3.13, 3.14, if only one edge is connected to the generic node v, then $\mathscr{C}_I(v) \cup \mathscr{C}_O(v) = \{i\}$. Hence, the sums $\sum f_k, \sum(\mu_k + f_k \times r_{k,C_v})$ (being k the generic index for the edge) are null and the equations are basically simpler. This case is peculiar, and its significance will be clear later on when the solution of the EOG is discussed in detail.

References

1. Cheli, F., & Pennestri, E. (2006). *Cinematica e Dinamica dei Sistemy Multibody*. Milano: Casa Editrice Ambrosiana.
2. Featherstone, R. (2010). Exploiting sparsity in operational-space dynamics. *The International Journal of Robotics Research*, *29*, 1353–1368.
3. Featherstone, R., & Orin, D. E. (2008). Dynamics. In B. Siciliano & O. Khatib (Eds.), *Springer Handbook of Robotics* (pp. 35–65). Berlin-Heidelberg: Springer.
4. Featherstone, R. (2007). *Rigid body dynamics algorithms*. Secaucus: Springer.

5. Fumagalli, M., Ivaldi, S., Randazzo, M., Natale, L., Metta, G., Sandini, G., et al. (2012). Force feedback exploiting tactile and proximal force/torque sensing—Theory and implementation on the humanoid robot iCub. *Autonomous Robots*, *33*, 381–398.
6. Ivaldi, S., Fumagalli, M., Randazzo, M., Nori, F., Metta, G., & Sandini, G. (2011). Computing robot internal/external wrenches by means of inertial, tactile and F/T sensors: Theory and implementation on the iCub. In *2011 11th IEEE-RAS International Conference on Humanoid Robots (Humanoids)*, (pp. 521–528).
7. Sciavicco, L., & Siciliano, B. (2005). *Modelling and control of robot manipulators* . Advanced textbooks in control and signal processing series(2nd ed.) London: Springer.
8. Sciavicco, L., & Siciliano, B. (2005). *Modelling and control of robot manipulators*. Advanced textbooks in control and signal processing. London: Springer.
9. Wittenburg, J. (1994). Topological description of articulated systems. *Computer-aided analysis of rigid and flexible mechanical systems* (pp. 159–196).

Chapter 4
Building EOG for Computing Dynamics and External Wrenches of the iCub Robot

Abstract Chapter 3 has shown a method that makes use of a graphical formulation employing graph theories for performing the computation of both kinematic quantities (i.e. angular velocities of the center of mass of links, but also linear and angular acceleration), and dynamic (internal forces and moment on the connection elements between the links, but also externally applied wrenches). This chapter shows the application of the method presented in Chap. 3, applied to the dynamic formulation of the iCub humanoid robot (see Chap. 2). Section 4.1 summarizes the steps required for building the kinematic and dynamic model of robotic mechanisms. The application of the method will be shown for different simple robotic structures, and will then be used for modeling the iCub robot. Results will be reported in Sect. 4.3.2, where the validation of the inverse dynamic algorithm and the model is addressed by means of the comparison between computed and measured internal wrenches at the sensor reference frame. Two other experiments show the comparison of externally applied forces measured with an external FTS, with virtual FTS exploiting the proposed method. Also a comparison of external torques on the joints will be performed.

4.1 Summary of the EOG Definition on Robotic Structure: Case Studies

Sections 3.4.1 and 3.4.2 presented the basic steps for propagating kinematic and dynamic information across a graph representing a kinematic tree (see also [4, 5]). This flow of information can be used to determine unknown (actually non directly measured) quantities along the kinematic tree. In particular, it is possible to have a *virtual measurement* of both kinematic quantities (ω_i, $\dot{\omega}_i$ and $\ddot{p}_i$ for each link i) and dynamic (f_i and μ_i).

The discussion addressed in this chapter will mainly focus on the *virtual force/torque sensor* method, exploiting the dynamical model of the robotic structure, to have an estimation of both internal wrenches that the links mutually exchange on

M. Fumagalli, *Increasing Perceptual Skills of Robots Through Proximal Force/Torque Sensors*, Springer Theses, DOI: 10.1007/978-3-319-01122-6_4,

joint connections, and externally applied generalized forces at any location of the kinematic structure. About the popagation of kinematic quantities, only the case in which one inertial sensor is present on the structure will be addressed. This is in fact sufficient to obtain an estimation of the kinematic quantities along all the structure. Multiple case studies, where one or more distributed FTSs are engaged, will be analyzed to better understand the propagation of wrenches within kinematic trees. Note that given N FTS distributed along the kinematic chain, $N+1$ *virtual measurement* of externally applied forces (one for each sub-chain) can be performed. Additionally the unknown leaves can be dynamically moved along the graphs, accordingly to the point of application of the externally applied wrenches. Therefore the solution of the graph results in a non fixed path followed by the information flow during the computation, as already mentioned in Sect. 3.3.

This section summarizes the basic steps to compute the whole-body dynamics, with specific attention at getting estimates for the externally applied wrenches (denoted with $\lozenge$).

Hereafter follows the steps for the definition of the graph structure:

1. Create the graph representing the kinematic tree; define a node for each link and an edge for each joint connecting two links. The edge orientation is arbitrary and in particular it can be defined according to a "regular numbering scheme".
2. For each inertial sensor (measuring the linear acceleration and the angular velocity and acceleration) insert a *black triangle* ($\blacktriangledown$) and an edge from the node v to the triangle, where v represents the link to which the sensor is attached. Associate to the edge the reference frame $\langle s \rangle$ corresponding to the sensor frame.
3. For any node v with only ingoing edges, add a *white triangle* ($\triangledown$) and an edge from v to the triangle. Associate to the edge an arbitrary reference frame $\langle v \rangle$.

At point 2, it should be noticed that kinematic chains are often grounded and therefore there exists a base link with null angular kinematics, $\omega = [0, 0, 0]^\top$, $\dot{\omega} = [0, 0, 0]^\top$ and gravitational linear acceleration $\ddot{p} = g$, being g the vector representing the gravity force (as an example, $g = [0, 0, -9.81]$ if the base frame has the z-axis oriented as the gravity component. Any other condition is obviously allowed, and depend on the orientation of the base frame).

Computations can be performed following the procedure in Algorithm 4.1, that is a *pre-order* traversal of the tree with elementary operations defined by Eqs. 3.8 and 3.9 or 3.10. If multiple $\blacktriangledown$ nodes (i.e. inertial sensors) are present in the graph, each path between two of these nodes corresponds to a set of three equations containing the measurements: one for the linear accelerations, one for the angular velocity and one for the angular accelerations. These equations can be used to refine the sensor measurements or to give better estimates of the joint velocities and accelerations (typically derived numerically from the encoders and therefore often noisy), as reported in Appendix A.

Figure 4.1 shows an example of a floating base multiple branches mechanism which mount one inertial sensor and multiple FTSs (see Fig. 4.1a). The kinematic of this chain can be defined starting from link 0 (Fig. 4.1b). The enhanced graph

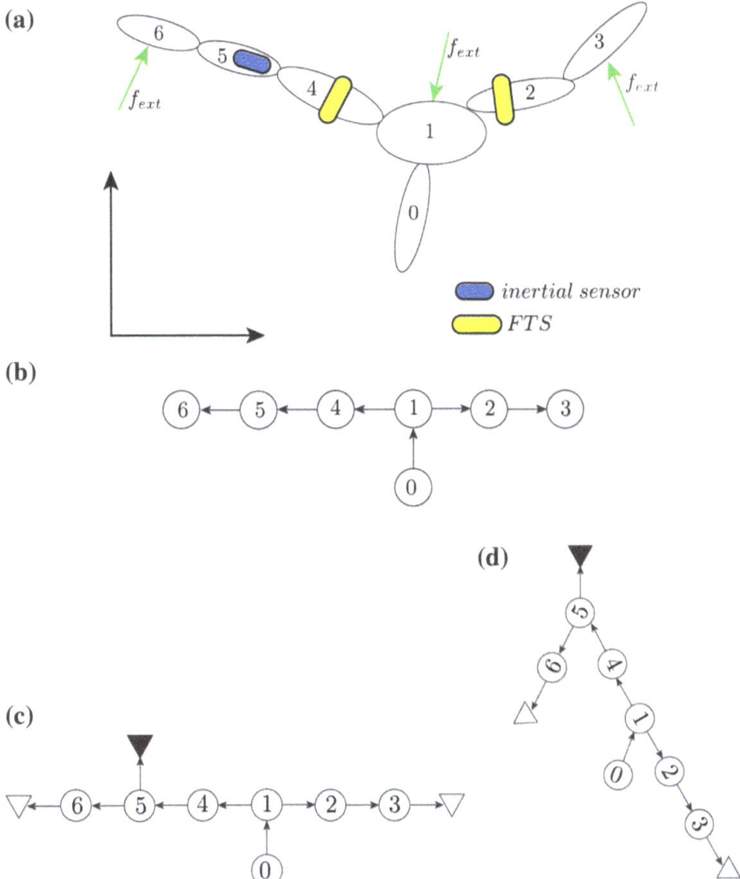

Fig. 4.1 **a** An example of generic floating, multiple branches kinematic chain subject to external forces; The chain mounts two FTSs and one inertial sensors. **b** Classical representation of the graph structure; The direction of the edges represents the kinematic conventions for the definition of the kinematic of the chain. **c** Enhanced graph representation of the kinematic of the chain. **d** Rearrangement of the kinematic EOG to underline the way to visit the graph, actually a pre-order traversal; starting from the root node, the computation is performed in the following order: 5, 6, unknown leaf, 4, 1, 0, 2, 3, unknown leaf

associated to this mechanism takes the form of the graph shown in Fig. 4.1c, which can be rearranged as in Fig. 4.1d. It is clear that the way to visit the graph is a pre-order traversal. It is remarkable here that the visit order is not related to the edge direction, since the latter only affects the recursive equations that must be used to propagate the variables, as was already shown with Fig. 3.10. Once velocities and accelerations have been computed for all edges, a new series of steps needs to be performed on the EOG to obtain the dynamic enhanced subgraphs.

1. For each FTS embedded in the link v cut the graph into two subgraphs according to the procedure shown in Fig. 3.8d. Divide v into two nodes v_{i,S_B} and v_{i,S_F} representing the sub-links (with suitable dynamic properties); define two *black rhombi* (♦) and add two edges from the rhombi to the nodes. Associate to both the edges the same reference frame $\langle s \rangle$ corresponding to the sensor frame.
2. If there are other known wrenches acting on a link (e.g. sensors attached at the end-effector), insert a *black rhombus* (♦) and an edge from the rhombus to v, where v represents the link to which the wrench is applied. Associate to the edge the reference frame $\langle s \rangle$ corresponding point where the external wrench is applied.
3. If the distributed tactile sensor [1] is detecting externally applied wrenches, insert a *white rhombus* (◊) for each externally applied unknown wrench. Add an edge connecting the rhombus with v, where v represents the link to which the wrench is applied. The edge orientation is arbitrary depending on the wrench to be computed (i.e. the wrench from the link to the external environment or the equal and opposite wrench from the environment to the link). Associate to the edge the reference frame $\langle c \rangle$ corresponding to the location where the external wrench is applied.

After these steps have been performed, the dynamic enhanced subgraphs are obtained, each of which can be considered independently. Wrenches can be propagated to the unknown nodes (◊) if and only if there exists a unique unknown for each sub-graph. If this is the case, then for each unknown we can define a tree with the node ◊ as root. Wrenches can be propagated from the leaves to the root following the procedure in Algorithm 4.2, which is basically a *post-order* traversal of a tree(see [2]) with elementary operations defined by Eqs. 3.13 or 3.14. Figure 4.2 shows the graph associated to the mechanism of Fig. 4.1a. In this case, since there are 2 FTSs in the chain, which correspond to 2 known quantities in the enhanced graph (see Fig. 4.2a), an overall of 3 external forces can be determined, which correspond to 3 unknown nodes in the enhanced graph. To perform the computation, the graph should therefore be splitted into 3 subgraphs (see Fig. 4.2b), each of which allows to detect a single external wrench ◊. If there is no ◊ node in a subgraph (i.e. no external forces are acting on the subgraph), then the *post-order* traversal of this graph produces two equations (one for forces and the other for wrenches) with no unknowns.[1] These equations can be used to estimate on-line the dynamical parameters of the corresponding kinematic sub-tree exploiting the linearity of these parameters in the equations (see [9]).

Remarkably, in the considered cases (one ◊ per subgraph at maximum) each edge in the subgraph is visited during the *post-order* traversal. As a result, all internal wrenches are computed and therefore a complete characterization of the whole-body dynamics is retrieved.

It is now clear that, as a consequence of what has been shown, given N FTS distributed on a chain, $N + 1$ sub-graphs are produced and therefore a maximum of $N + 1$ external wrenches can be estimated (one for each sub-graph).

[1] Practically, these equations can be obtained by defining an arbitrary ◊ connected to an arbitrary node. A *post-order* traversal of the graph with ◊ as root determines the equations by simply assuming that the wrench associated to the edge connected to ◊ is null.

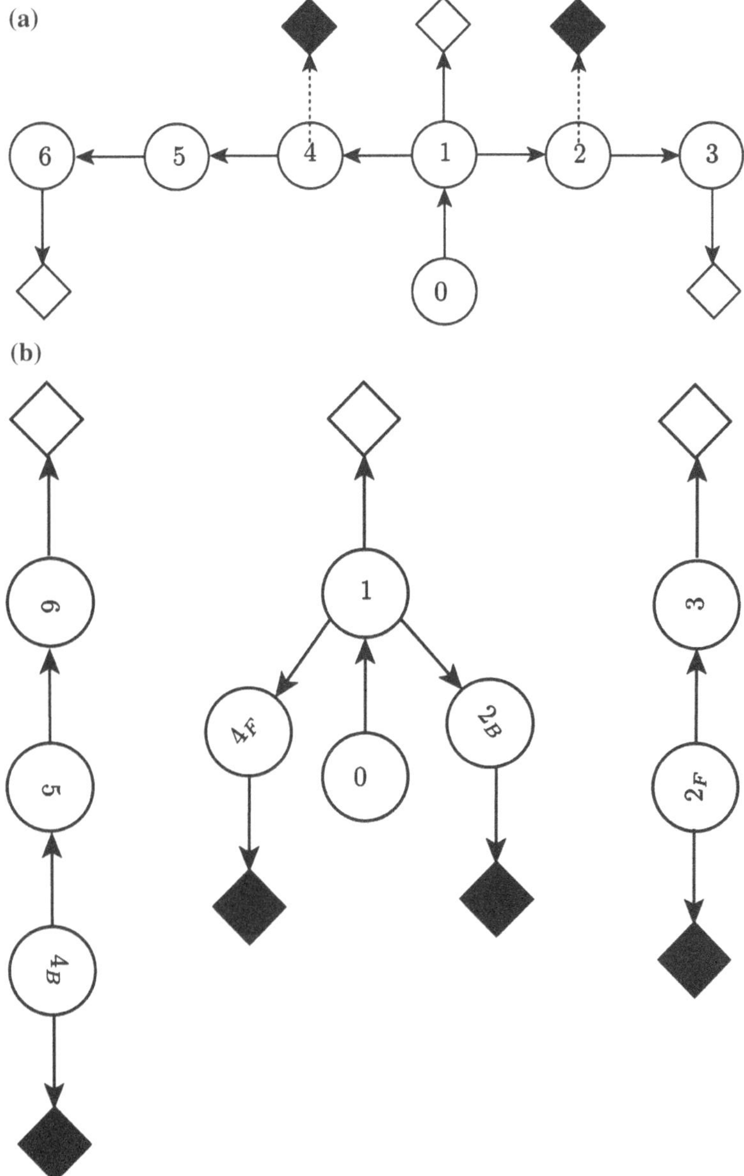

Fig. 4.2 The figure refers to the floating, multiple branches kinematic chain subject to external forces shown in Fig. 4.1a. **a** Enhanced graph representation of the kinematic of the chain. Two known quantities are present (♦), which represent the FTSs, together with the three unknowns representing the external forces (◇); nodes 4 and 2 will be divided into two sub-nodes, giving origin to three sub-chains. **b** Rearrangement of the dynamic EOGs to underline the way to visit the graph, actually a post-order traversal; starting from the leaf nodes, the computation is performed in the following order: (case 1(*left*)) 4_B, 5, 6, unknown leaf, (case 2(*center*)) 4_F, 0, 2_B, 1, unknown leaf, 2_F, 3, unknown leaf

Algorithm 4.1 Solution of kinematic EOG exploiting a tree

Require: EOG, $\omega_0, \dot{\omega}_0, \ddot{p}_0$
Ensure: $\omega_i, \dot{\omega}_i, \ddot{p}_i, \forall v_i$
Attach a node ▼ for every kinematic source (e.g. inertial sensor)
Set $\omega_0, \dot{\omega}_0, \ddot{p}_0$ in ▼
Re-arrange the graph with a ▼ as the root of a tree
KinVisit(EOG,v_{root})

KinVisit(EOG,v_i)
Compute $\omega_i, \dot{\omega}_i, \ddot{p}_i$ with Eqs. 3.8 or 3.9 or 3.10 according to direction of the edges i, j connected to v
for all child v_k of v_i **do**
    KinVisit(EOG,v_k)
end for

Algorithm 4.2 Solution of dynamic EOG exploiting a tree

Require: EOG, $w_s \forall$ FTS
Ensure: $w_i, \forall v_i$
For every FTS, attach a node ◆ to the corresponding link
Set w_s in each ◆
For each ◆, split the graph and create two sub-graphs (see text for details)
Attach a node ◇ to each link where a contact is detected: if there is no contact in a subgraph, choose an arbitrary position and attach a fictitious ◇[2]
Re-arrange each sub-graph with a ◇ as the root of a tree
for all subgraph **do**
    DynVisit(EOG,v_{root})
end for

DynVisit(EOG,v)
if v has children **then**
    for all child $e_{v,h} \in \mathscr{C}(v)$, $e_{v,h} \neq i$ **do**
        $w_{e_{v,h}} = DynVisit(\text{EOG},h)$
    end for
end if
Compute w_i with Eqs. 3.13 or Eqs. 3.14 according to the direction of the edges

4.2 Performing the Computation

With reference to Fig. 4.2, different situation are reported hereafter. Note that the use of the Denavit-Hartenberg notation for the definition of the kinematic structure of the links, and the RNEA for the definition of the dynamics of the system, are not mandatory. Custom choices can be adopted.[2]

When propagating the wrenches from one vertex to the other, Eqs. 3.13 or 3.14 is used. This equation provides the computation of both internal and external forces, depending on the definition of known and unknown variables on the graph. With reference to Fig. 4.2b, when the flow of the information is along the same direction

[2] In this cases, usually the terminal link or the end-effector is selected.

of the edge, f_i of Eqs. 3.13 or 3.14 is to be computed. One of the forces among the $\sum f_k$ otherwise, depending on which of the k links the unknown is located.

4.2.1 Single-branched Open Chain

Refering to Fig. 4.2b, left and right graphs represent the extremities of a chain that have been slit at the level of the FTSs into two single branch open chains. In this situation, when the flow of the information is along the same direction of the edge, term f_i of Eq. 3.13 is to be determined; term f_i of Eq. 3.14 otherwise. Next sections show the case which demonstrate the generality of the EOG method also for open, multi-branched kinematic chains.

4.2.2 Multiple-branched Nodes and External Forces

With respect to Fig. 4.2b, we point out that the unknown $\diamondsuit$ attached to the base is used if a contact is detected on that link (e.g. if the artificial tactile skin reveals a contact at the base). In absence of contact (as in the case of node 0), the node $\diamondsuit$ is not needed. More in general, if f_{ext} is not present, it is possible to write the recursive equations as a compact set, where all the dynamic variables are known: this formulation can be exploited to obtain, for example, a better estimate of the rigid-body model parameters, e.g. links mass.

On the other hand, external forces may be acting in other locations different from the end-effector (e.g. on an internal link in between the base and the end-effector), as a consequence of contacts with the environment. In such cases, the application point (or the centroid of the contact region) must be known [1]. Note that one external force can be determined if, and only if, all the other wrenches flowing through the edges connected to the link can be determined.

Consider the general example of one link connected to N other links, $N \geq 2$ (i.e. node 1 of Fig. 4.2). The graph associated to a similar situation instead is the central one of Fig. 4.2b. The first step consists in setting the unknown wrenches given the quantities that have flown from the known leaves. These quantities can in general be measured by FTS within a link. Secondly each of the links 4_F, 0 and 2_B connected to 1 perform the calculation (using Eqs. 3.13 or 3.14) necessary to define the information passing through the edge which connect them to link 1, according to the direction of the edge. Then vertex 1 preforms again the evaluation of the force transmitted through the edges connecting 4_F, 0 and 2_B to 1, to perform the computation of the quantities flowing through the unknown edge, again from Eqs. 3.13 or 3.14, according with the direction of the edge $e_{j,i}$. Note that in this example, the assumption that f_{ext} is the only unknown must hold.

4.2.3 Virtual Joint Torque Sensors

In case the Denavit–Hartenberg notation is used for the definition of the kinematic of the structure, an estimation of the joint torque can be performed, once the i-th wrench is known, following the equation:

$$\tau_i = \mu_i^\top z_{i-1} \tag{4.1}$$

where z_{i-1} is the z-axis of the reference frame $\langle i-1 \rangle$ as in Fig. 3.3 (see [9]). The method shows that it is possible to have an estimation of joint torques, which can be used for joint torque control. Moreover, it must be noted that joint torques are derived as one component of the wrenches flowing through the edges. These wrenches allow having a better representation of the possible contact situation, which can be used as a virtual measurement, to perform every kind of tasks involving force detection and control. Therefore, the more 6-axis FTS are employed, the more accurate will be the estimation [4, 5, 8].

4.3 A Case Study: iCub Dynamics

The method described in Sect. 3.3, has been implemented on the 53 DoF humanoid robot iCub. The hands have been considered as a unique rigid body, assuming that the motion of the fingers does not contribute to the variation of FTSs measurements. The same assumption has been considered for the eyes. The resulting overall number of degrees of freedom of the dynamical model of the robot is 32: 7 for each of the two arms, 6 for the legs, 3 for the torso and 3 for the neck. This section shows the graph which refers to the kinematic and dynamic of the iCub. Its sources of information are reported in Sect. 4.3.1, while Sect. 4.3.2 show the results obtained which validate the dynamical model necessary to improve the perceptual capabilities of the humanoid platform and consequently design force control for the iCub robot.

4.3.1 Sensors

As presented in Chap. 2 , iCub is equipped with one inertial sensor (Xsens MTx-28A33G25 [10]) at the top of the head (see Fig. 4.3), and four custom-made 6-axes FTSs (see [3]), one per leg and arm, each placed proximally. Excluding the hands DOF from the model, 32 DOF have been taken into account. The structure of the enhanced graphs are reported and described in Sects. 4.3.1.1 and 4.3.1.2.

This section gives a detailed description of the information that can be retrieved on the iCub robot.

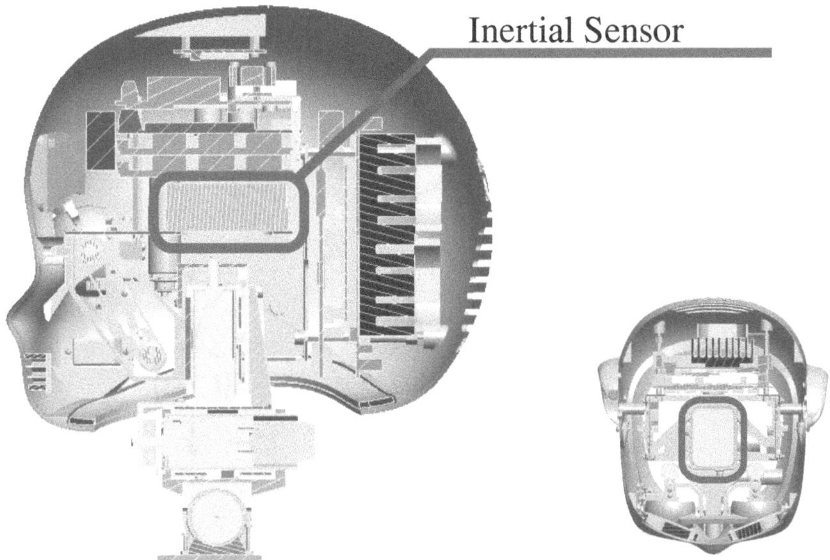

Fig. 4.3 Section of the iCub head, to show once more the position of the 3 DOF Orientation Tracker

4.3.1.1 The Inertial Sensor

The inertial sensor provides angular velocities and linear and angular acceleration at the reference frame of the sensor. When connected to a link, these quantities can be propagated to the entire link (see Eq. 3.10), or to the connected links (with Eqs. 3.9 or 3.8) depending on the direction of the edges that connect the nodes. The graph associated to the propagation of these information is shown in Fig. 4.4. The structure of the enhanced graphs which is required to perform the computation of the robot kinematic quantities has one single known input (▼) placed at the final node of the head. In order to perform the computation of the wrenches along the structure, all the nodes have to be visited in this phase. Moreover, given the kinematic representation of the end-points of the limbs, kinematic unknowns should be added (▽). Their edges represent the frame of reference of the links they are connected to. It is noticeable that, in case one or more encoder measurements are missing, a single inertial sensor would not be sufficient to propagate the information along the entire kinematic chain. On the other hand, multiple inertial sensors might be employed to perform an improved estimation of the joint velocities and acceleration, as reported in Appendix A.

As a final remark, it is to be underlined that the inertial sensor allows to perform the Newton–Euler computations without a fixed base frame (as it is usually assumed in its classical applications). A clarifying example can be the case in which the iCub crawls as shown in Fig. 4.5

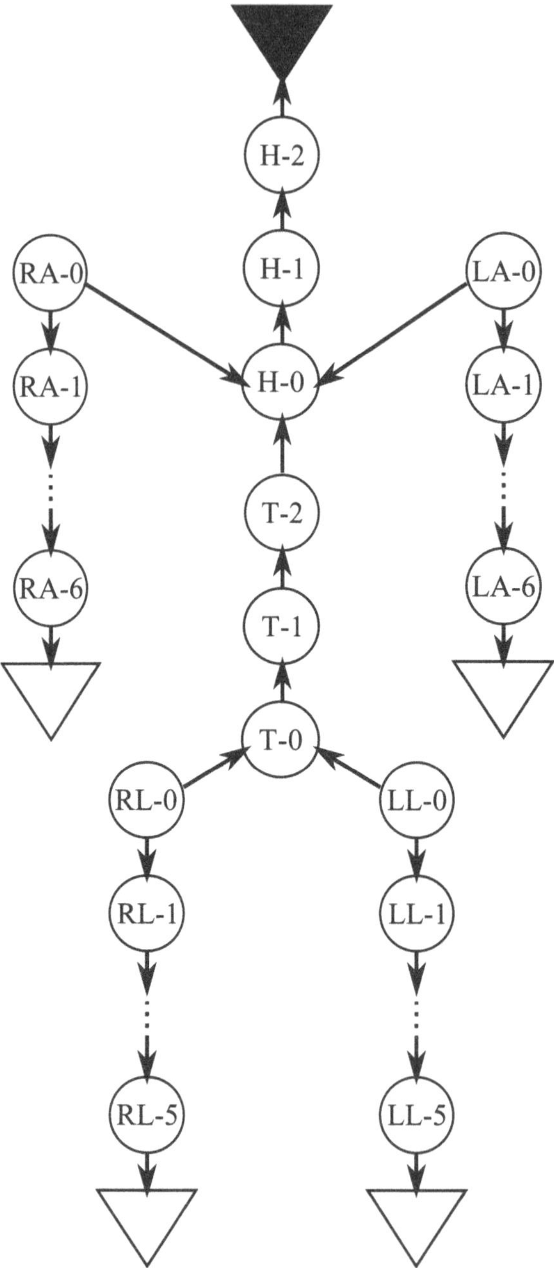

Fig. 4.4 Representation of iCub's kinematic graph, using the notation of Fig. 3.6. A complete description of the iCub kinematics can be found in the online documentation available on the iCub website wiki, at the page: http://eris.liralab.it/wiki/ICubForwardKinematics

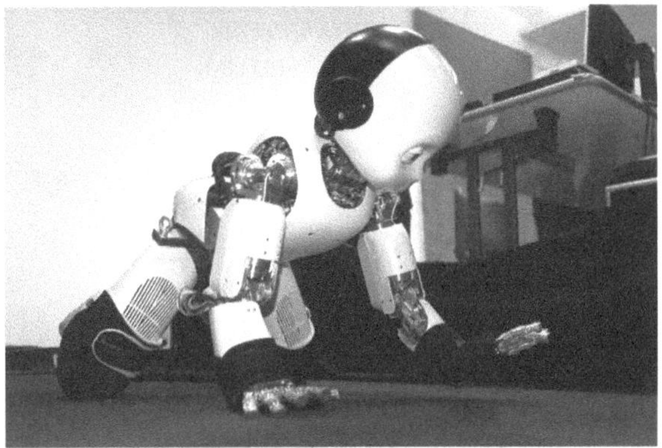

Fig. 4.5 The humanoid robot iCub performing the crawling task. The task stimulates all the sensors, from the externally applied reaction forces of the floor

4.3.1.2 The Force/Torque Sensor

A sensor embedded in a link (see Fig. 4.6) allows to divide the link into two sub-links, where the equilibrium is guaranteed by assigning to the edge, connecting the sub-links, the measure of the FTS. Practically it correspond to dividing the graph into two sub-graphs and introducing two black rhombi (i.e. two known wrenches), one on each sub-graph. More specifically, the sensor measures the wrench exerted by the "forward" sub-link to the "backward" sub-link (this will be represented by a first rhomboidal node). However, a wrench equal and opposite to the sensor measurement is also exerted by the "backward" sub-link to the "forward" sub-link (this will be represented by a second rhomboidal node). Under these considerations, the FTS within a link will be represented by splitting the node associated to the link into two sub-nodes (with suitable dynamical properties, see Fig. 4.6). On each sub-node, the known applied wrench will be represented with black rhomboidal nodes.

Figure 4.9 shows the graph corresponding to the iCub dynamic model. As represented in Figs. 4.7 and 4.8, the iCub robot mounts a set of four distributed FTSs [7]. Each FTS is placed in a proximal position within the limb. The FTSs of the arms are placed right after the 3-DOF shoulder universal joint, while the FTSs of the legs are

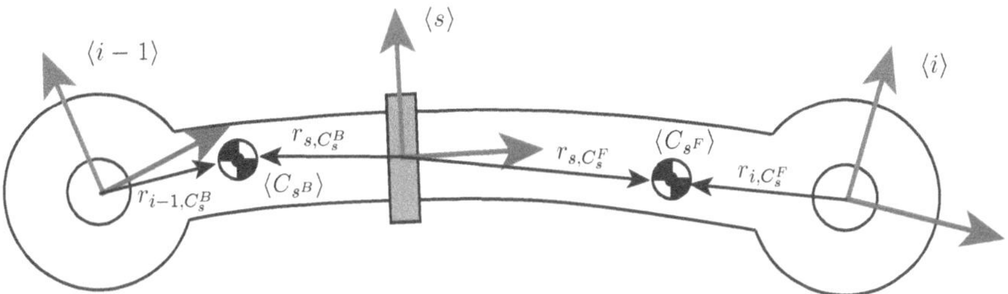

Fig. 4.6 A representation of an F/T Sensor within the i_S-th link. Note that the sensor divide the link into two sub-links, each with its own dynamical properties

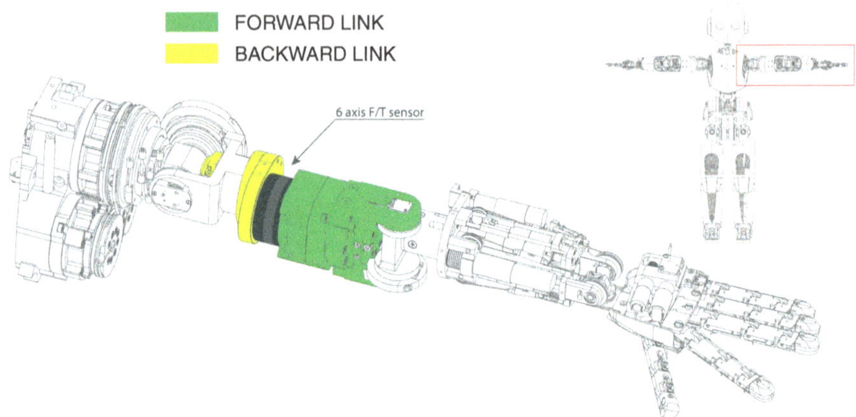

Fig. 4.7 The iCub arm. A CAD view of the iCub arm to put in evidence the presence and the position of the F/T sensor

placed after the first 2 joints of the hip. The corresponding graph can thus be divided into 5 sub-chains. Four chains are serial, single branched kinematic chain, while one is a multiple branched kinematic tree. The resulting graphs show that it is possible to detect a total of 5$\Diamond$, one for each sub-chain.

4.3.2 Experiments

The aim of this section is to validate the theoretical methodology presented in Chap. 3. Experiments have been conducted on the iCub robot. A dynamical model with the form of the graphs represented in Figs. 4.4 and 4.9 has been built according to the kinematic structure of the iCub robot. A 3D Orientation Tracker Xsens MTx placed on its head allows to measure ω, $\dot{\omega}$, $\ddot{p}$ for the head link. Encoders are used to measure all the joints positions and joint velocities and accelerations are derived from position measurements through a least-squares algorithm based on an adaptive window [6]. Force/torque sensors mounted proximally in the limbs (see Figs. 4.7 and 4.8) also allow to measure the forces acting in between the sensor and distal joints.

Three experiments show the effectiveness of the proposed methodology. In the first experiment, the validation of the dynamical model is performed by comparing measurements from the FTSs with their prediction based on the sole dynamical model. In this experiment the limbs move freely, without interaction or externally allpied forces. In the second experiment a commercial FTS is used as a tool to produce a known external wrench on the robot and to compare the external sensor measurement with the external wrench computation obtained as described in Sect. 4.1. Finally, we tested our procedure for computing joint torques (4.1) by comparing our joint torque

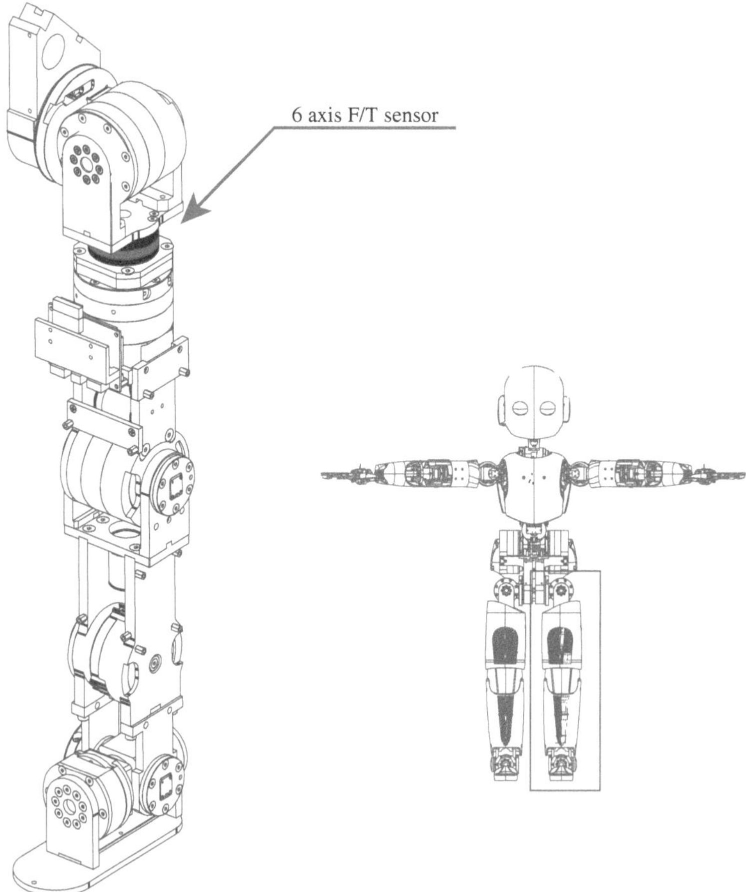

Fig. 4.8 The iCub leg. A CAD view of the iCub leg to put in evidence the presence and the position of the F/T sensor

estimation with a joint torque measurement obtained by projecting a known wrench (once again measured with an external sensor) on the joints.

4.3.2.1 Validation of the Dynamical Model

In a first experiment, the validity of dynamical model of the iCub limbs is tested. The measurements w_s of the four 6-axes FTSs have been compared with the quantity $\hat{w}_s$ predicted by the dynamical model. Sensor measurements w_s can be predicted assuming null wrench at the limbs extremities (hands or feet) and then propagating forces up to the sensors. Data presented in this section were recorded under this assumption. Table 4.1 summarizes the statistics of the errors $w_s - \hat{w}_s$ for each limb

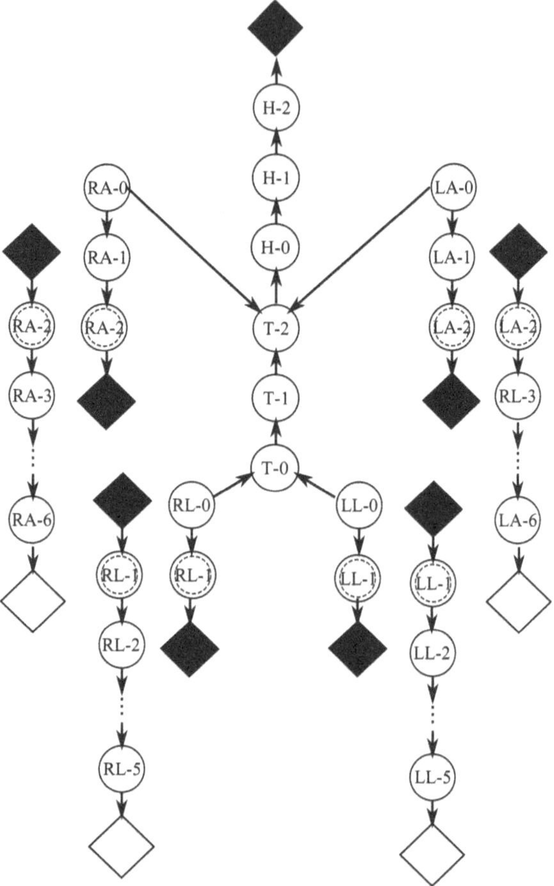

Fig. 4.9 Representation of iCub's dynamic enhanced graph. Notice that all the limbs are divided into two sub-graphs in correspondence of the the F/T sensors located as sketched in Figs. 4.7 and 4.8

during a sequence of movements. In particular, the table shows the mean and the standard deviation of the errors between measured and predicted sensor wrench during the movements. Figure 4.10 plots w_s and $\hat{w}_s$ for the left arm during the same sequence of movements.

4.3.2.2 Estimation of External Wrench

In the second experiment, the effectiveness of procedure proposed for measuring unknown external wrenches, described in Sect. 4.1, is verified. The "unknown" wrenches are generated with the help of an external 6-axes FTSs that give a ground truth of the applied wrench. An external wrench w^E was applied at the left hand and measured with the external F/T sensor. Its value was then compared with the estimation of the external wrench $\hat{w}^E$, obtained by propagating the internal FTS

Table 4.1 Errors in predicting F/T measurement (see text for details)

	right arm: $\varepsilon \triangleq \hat{w}_{s,RA} - w_{s,RA}$					
	ε_{f_0}	ε_{f_1}	ε_{f_2}	ε_{μ_0}	ε_{μ_1}	ε_{μ_2}
$\bar{\varepsilon}$	−0.3157	−0.5209	0.7723	−0.0252	0.0582	0.0197
σ_ε	0.5845	0.7156	0.7550	0.0882	0.0688	0.0364
	left arm: $\varepsilon \triangleq \hat{w}_{s,LA} - w_{s,LA}$					
	ε_{f_0}	ε_{f_1}	ε_{f_2}	ε_{μ_0}	ε_{μ_1}	ε_{μ_2}
$\bar{\varepsilon}$	−0.0908	−0.4811	0.8699	0.0436	0.0382	0.0030
σ_ε	0.5742	0.6677	0.7920	0.1048	0.0702	0.0332
	right leg: $\varepsilon \triangleq \hat{w}_{s,RL} - w_{s,RL}$					
	ε_{f_0}	ε_{f_1}	ε_{f_2}	ε_{μ_0}	ε_{μ_1}	ε_{μ_2}
$\bar{\varepsilon}$	−1.6678	3.4476	−1.5505	0.4050	−0.7340	0.0171
σ_ε	3.3146	2.7039	1.7996	0.3423	0.7141	0.0771
	left leg: $\varepsilon \triangleq \hat{w}_{s,LL} - w_{s,LL}$					
	ε_{f_0}	ε_{f_1}	ε_{f_2}	ε_{μ_0}	ε_{μ_1}	ε_{μ_2}
$\bar{\varepsilon}$	0.2941	−5.1476	−1.9459	−0.3084	−0.8399	0.0270
σ_ε	1.8031	1.8327	2.3490	0.3365	0.8348	0.0498

$\varepsilon \triangleq \hat{w} - w = [\varepsilon_{f_0}, \varepsilon_{f_1}, \varepsilon_{f_2}, \varepsilon_{\mu_0}, \varepsilon_{\mu_1}, \varepsilon_{\mu_2}]$; SI Unit: $f : [N]$, $\mu : [Nm]$

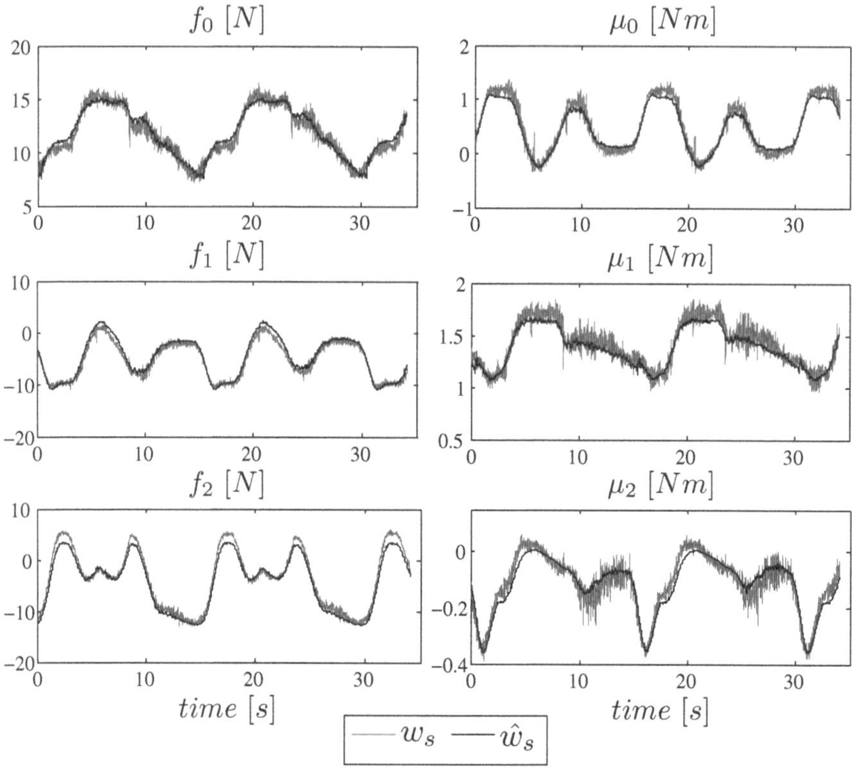

Fig. 4.10 Left arm: comparison between the wrench measured by the FT sensor and the one predicted with the model, during the "Yoga" demo

measurements from the left arm sensor to the frame where w^E was applied. The propagation was performed according to th enhanced graph of Fig. 4.4 representing the left arm. A plot of w^E and $\hat{w}^E$ for different values of the stimulus is given in Fig. 4.11.

4.3.2.3 Estimation of External Torques

A third experiment was performed to test the validity of the proposed procedure for estimating joint torques from the embedded F/T sensor.[3] Let this estimation be called $\hat{\tau}$. In this case, given the difficulties in generating known torques at the joints, we proceeded as in the previous experiment generating a wrench w^E and computing the corresponding torques τ^E at the joints with the following formula:

---

[3] Torques estimation from embedded F/T sensors can be obtained using (4.1) with the μ_i computed propagating the F/T sensor information within the graph according to the procedure presented in previous sections.

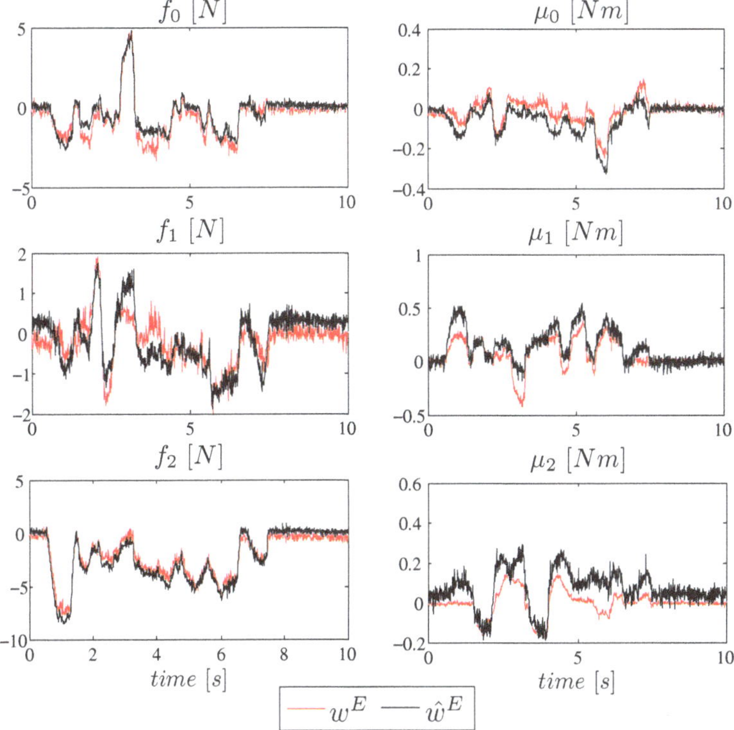

Fig. 4.11 Left arm: comparison between the external wrench estimated after the FT sensor measurements and the one measured by an external FT sensor, placed on the palm of the left hand

$$\tau^E = J_E^\top w^E, \tag{4.2}$$

where $J_E \in \mathbb{R}^{6\times n}$ is the Jacobian for the given wrench application point. The torques τ at the joints generally differ from τ^E since we have $\tau = \tau^I + \tau^E$, where τ^I represents the vector of joint torques due to the system internal dynamics. Keeping the robot fixed and assuming w^E null, we have $\tau = \tau^I$ and we can use $\hat{\tau}$ to obtain an estimation of τ^I, denoted $\hat{\tau}^I$. Keeping the robot in the same configuration but letting $w^E \neq 0$, we can then estimate τ^E with the following formula:

$$\hat{\tau}^E = \hat{\tau} - \hat{\tau}^I.$$

Figure 4.12 shows a comparison between τ^E and $\hat{\tau}^E$ obtained with the procedure just described.

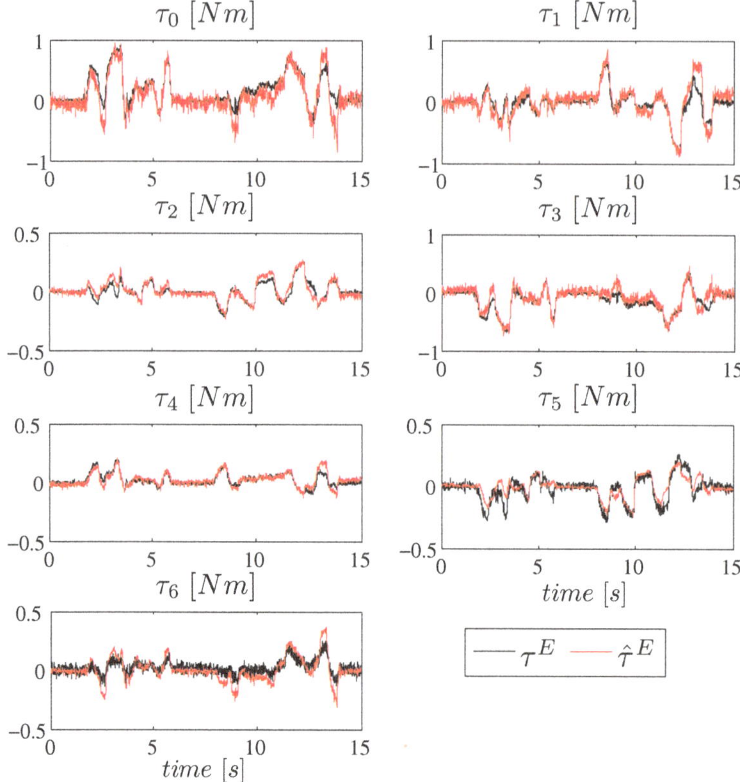

Fig. 4.12 Left arm: comparison between the torques computed exploiting the FT sensor and the ones obtained by projecting the external FT sensor on the joints

References

1. Cannata, G., Maggiali, M., Metta, G., & Sandini, G. (2008). An embedded artificial skin for humanoid robots. In *IEEE International Conference on Multisensor Fusion and Integration*. Seoul, Korea.
2. Cormen, T., Leiserson, C., Rivest, R., & Stein, C. (2001). *Introduction to algorithms* (2nd ed.). Cambridge (Massachussets): the MIT Press and McGraw Hill.
3. Fumagalli, M., Randazzo, M., Nori, F., Natale, L., Metta, G., & Sandini, G. (2010). Exploiting proximal F/T measurements for the iCub active compliance. In *IEEE/RSJ International Conference on Intelligent Robots and Systems*, Taipei, Taiwan.
4. Fumagalli, M., Ivaldi, S., Randazzo, M., Natale, L., Metta, G., Sandini, G., et al. (2012). Force feedback exploiting tactile and proximal force/torque sensing—Theory and implementation on the humanoid robot iCub. *Autonomous Robots, 33*, 381–398.
5. Ivaldi, S., Fumagalli, M., Randazzo, M., Nori, F., Metta, G. & Sandini, G. (2011). Computing robot internal/external wrenches by means of inertial, tactile and F/T sensors: Theory and implementation on the iCub. In *Humanoid Robots (Humanoids), 11th IEEE-RAS International Conference on 2011* (pp. 521–528).

6. Janabi-Sharifi, F., Hayward, V., & Chen, C. S. J. (2000). Discrete-time adaptive windowing for velocity estimation. *IEEE Transactions on Control Systems Technology*, *8*, 1003–1009.
7. Natale, L., Nori, F., Metta, G., Fumagalli, M., Ivaldi, S., Pattacini, U., et al. (2012). The iCub platform: a tool for studying intrinsically motivated learning. In Baldassarre, G., & Mirolli, M. (eds.), *Intrinsically motivated learning in natural and artificial systems*. Heidelberg: Springer-Verlag.
8. Randazzo, M., Fumagalli, M., Nori, F., Natale, L., Metta, G., & Sandini, G. (2011). A comparison between joint level torque sensing and proximal F/T sensor torque estimation: Implementation on the iCub. In *Humanoid Robots (Humanoids), Intelligent Robots and Systems (IROS), IEEE/RSJ International Conference on 2011*, (pp. 4161–4167).
9. Sciavicco, L., & Siciliano, B. (2005). Modelling and control of robot manipulators. In *Advanced textbooks in control and signal processing*. Heidelberg: Springer.
10. XsensMTx (2010). The MTx orientation tracker user manual. http://www.xsens.com/en/general/mtx

Chapter 5
Active Compliance Control

Abstract The iCub robot is generally constituted by particular types of mechanisms where every single motor do not directly actuate one coresponding joint. The main example is the shoulder transmission of the iCub robot. This mechanism is constituted by three motors with parallel axis moving idle pulleys that cooperate together to move a universal 3DoF joints. This slution allows compactness and wide range of movements, similar to that of human shoulder. An advantage is that the overall moving mechanism results lightweight because of the allocation of the motors in the torso. On the other side, the control of such mechanisms requires to be analyzed. In this chapter, an analysis of the dynamic of coupled transmissions is conducted. The chapter highlights the problematics related to the motion control of coupled mechanisms, and addresses the derivation of control strategies for interaction conrol.

5.1 Dynamics of Coupled Mechanism

A coupled mechanism is a mechanical transmission which connects the motors to the joints by means of kinematic relationships. Generally speaking, a transmission mechanism allows the transer of power between the motor side of the mechanism, and the joint side (also called load). The dynamics of motors and joints cannot therefore be studied separaely.On one hand, the motors should cooperate to perform pure joint movements. On the other hand, moving one joint means transfering power to more than one motor. In this Section, the dynamics of the complete system is derived. Section 5.1.1 shows the dynamic of the motors and of the joints separately, while in Sect. 5.1.2 the equations of the transmission are defined, thus giving origin to the overall system dynamic equations presnted in Sect. 5.1.3. Control strategies will finally be derived, taking into consideration the dynamics of the coupled mechanism (see Sect. 5.2).

M. Fumagalli, *Increasing Perceptual Skills of Robots Through Proximal Force/Torque Sensors*, Springer Theses, DOI: 10.1007/978-3-319-01122-6_5,

5.1.1 Dynamic of Motors and Links

This paragraph reports the dynamics of motors and joints. The two sides of the transmission are here considered separately. The dynamic of the motors can generally be described by means of its mechanical and electrical equations. The mechanical equation representing the motors' dynamics is reported in Eq. (5.1):

$$I_m\ddot{\theta}_m + D_m\dot{\theta}_m = \tau_m - \tau_{mj} \tag{5.1}$$

where I_m and $D_m \in \mathbb{R}^{N\times N}$ represent the diagonal inertia and damping matrices of N motors, $\tau_m \in \mathbb{R}^N$ is the vector of motor torques, while term $\tau_{mj} \in \mathbb{R}^N$ represents the torque vector which the joint side of the mechanism ransfers to the motor side. In other words, this term represent the overall joint torque that is reflected on the motor side through the transmission mechanism. The mechanical and electrical dynamic are linked together through electro mechanical interactions. Let us generally consider the electrical dynamic of the motors as:

$$L_m\dot{i} + R_m i = V_{in} - V_{bemf} \tag{5.2}$$

being $L_m, R_m \in \mathbb{R}^{N\times N}$ the motor inductance and resistance of the spires of the motor and $V_{bemf} = k_\omega\dot{\theta}_m \in \mathbb{R}^N$ represents th *Back Electro-Motive Force* due to the relative motion of the magnetic field and the spires of the motor.

The overall torque that the magnetic field can generate on the motor shaft can be derived, for this kind of model as:

$$\tau_m = k_i i \tag{5.3}$$

being $i \in \mathbb{R}^N$ the current passing through the spires of the motors and $k_i \in \mathbb{R}^{N\times N}$ the torque constant diagonal matrix. The knowledge of the electrical dynamics of the motor is necessary for determining the current which flows into the wires, given an input voltage $V_{in} \in \mathbb{R}^N$.

The dynamics of the joint side of the transmission can be described, as already discussed in Chap. 3 by the nonlinear Eq. (5.4). Note that the input torque is not here the motor torque τ_m, but it is the motor torque that is reflected on the joint side $\tau_{jm} \in \mathbb{R}^N$. The joint equation of motion can therefore be written as:

$$I_j(\theta_j)\ddot{\theta}_j + C_j(\theta_j, \dot{\theta}_j)\dot{\theta}_j + G_j(\theta_j) = \tau_{jm} - \tau_j \tag{5.4}$$

where $I_j(\theta_j) \in \mathbb{R}^{N\times N}$, $C_j(\theta_j, \dot{\theta}_j)$ and $G_j(\theta_j) \in \mathbb{R}^N$ represent the Inertia matrix of the manipulator and the centrifugal and Corioli's term and gravitational contribution to the transmission system espectively, while $\tau_j \in \mathbb{R}^N$ is the external joint torque vector. For ease of treatment, the gravitational and centrifugal terms will be hereinafter included in a single term τ_j, which also includes the external joint torque components, i.e. external interaction torques. The overall equation of motion characterizing the joint side of the transmission thus becomes:

$$I_j(\theta_j)\ddot{\theta}_j = \tau_{jm} - \tau_j \tag{5.5}$$

5.1.2 Kineto-Static Equation of the Transmission

A mechanical transmission couples the dynamic of the joint side to the dynamics of the motors. In this treatment only rigid transmissions are considered. This assumption reduces the complexity of the treatment which derives from the introduction of elastic elements in the dynamic model of the robotic system. Linear and non dissipative transmission are therefore here considered. This allows to assume that the power exchange between the joint side P_j and the motor side P_m is not afected by energetic losses.

The kinematic of transmission mechanisms can be described as:

$$\theta_j = f(\theta_m) \tag{5.6}$$

being $f(.)$ a general function which describes the relationship between the motor coordinates θ_m and the joint θ_j. Deriving Eq. (5.6) with respect to time follows that the kinematic coupling between motor angular velocities $\dot{\theta}_m$ and joint velocities $\dot{\theta}_j$ can be described with a linear operator of the form:

$$\dot{\theta}_j = T_{jm}(\theta_m)\dot{\theta}_m \tag{5.7}$$

where $T_{jm}(\theta_j) \in \mathbb{R}^{n\times n}$ represents in general a non-linear operator (a Jacobian) which relates the velocities in the motor space $\dot{\theta}_m \in \mathbb{R}^n$ to the velocities of the joint space $\dot{\theta}_j \in \mathbb{R}^n$. Note that we are here considering fully actuated transmissions, which implies that the number of motors and the number of joints is the same. The above relationship represents the contribution of motor velocities to joint velocities. The inverse relationship of Eq. (5.6) can be derived as follows:

$$\theta_m = g(\theta_j) \tag{5.8}$$

that, deriving with respect to time, becomes:

$$\dot{\theta}_m = T_{mj}(\theta_j)\dot{\theta}_j \tag{5.9}$$

where $T_{mj}(\theta_j)$ is invertible and whose inverse is:

$$T_{mj}^{-1}(\theta_j) = T_{jm}(\theta_m) \tag{5.10}$$

If we now take into consideration the work δW_m produced by the motor side, which is transmitted to the joint side and the work δW_j produced by the load side, which is transmitted to the motors:

$$\begin{aligned} \delta W_m &= \tau_m j^\top \delta\theta_m = \tau_m j^\top \dot{\theta}_m \delta t \\ \delta W_j &= \tau_j m^\top \delta\theta_j = \tau_j m^\top \dot{\theta}_j \delta t \end{aligned} \tag{5.11}$$

In the hypothesis that the transmission is not dissipative and rigid, the overall instantaneous power $P\delta t$ (or virtual work δW, being $\delta W = P\delta t$) is conserved, which means that $\delta W_m = \delta W_j$. Therefore:

$$\tau_{mj}^\top \dot{\theta}_m = \tau_{jm}^\top \dot{\theta}_j \tag{5.12}$$

Let us now substitute (5.9) into (5.12)

$$\tau_{mj}^\top T_{mj}(\theta_j)\dot{\theta}_j = \tau_{jm}^\top \dot{\theta}_j \tag{5.13}$$

which is true for every non zero joint velocities if:

$$T_{mj}(\theta_j)^\top \tau_{mj} = \tau_{jm} \tag{5.14}$$

The inverse relationship is given by:

$$\tau_{mj} = T_{mj}(\theta_j)^{-\top} \tau_{jm} = T_{jm}(\theta_m)^\top \tau_{jm} \tag{5.15}$$

where $A^{-\top}$ indicates the transposed (pseudo-)inverse of a matrix A.

5.1.3 Dynamics

Given the consideration of the previous paragraph, the dynamic equations of the coupled mechanism can be derived as:

$$\begin{aligned} I_m\ddot{\theta}_m + D_m\dot{\theta} &= \tau_m - \tau_{mj} \\ I_j\ddot{\theta}_j &= n^\top T_{jm}^{-\top} \tau_{mj} - \tau_j \end{aligned} \tag{5.16}$$

where $n \in \mathbb{R}^{n \times n}$ is the diagonal matrix of reduction ratio of the gear boxes. T_{jm} represents the coupling matrix of the transmission, as presented in Eq. (5.7).

Combining τ_{mj} and considering Eq. (5.10), it is possible to express the dynamics of the transmission as:

$$I_m\ddot{\theta}_m + D_m\dot{\theta}_m = \tau_m - T_{jm}^\top n^{-\top}(\tau_j + I_j n^{-1} T_{jm}\ddot{\theta}_m) \tag{5.17}$$

which can be rearranged as:

$$\left(I_m + T_{jm}^{\top} n^{-\top} I_j n^{-1} T_{jm}\right) \ddot{\theta}_m + D_m \dot{\theta}_m = \tau_m - T_{jm}^{\top} n^{-\top} \tau_j \tag{5.18}$$

Equation (5.18) shows the dependency of the coupled system dynamic from both the motor torque and the external torque acting on the joint side. The resulting system dynamics thus become:

$$\tau_m = \left(I_m + \left(T_{jm}^{\top} n^{-\top} I_j n^{-1} T_{jm}\right)\right) \ddot{\theta}_m + D_m \dot{\theta}_m + T_{jm}^{\top} n^{-\top} \tau_j \tag{5.19}$$

5.2 Control

The goal of this section is to define a proper control strategy which allows to control pure joint movements, by assigning a control input to the motors. Introducing the electro-mechanical elationships of the motors, the dynamic of the transmission can be rewritten in the form:

$$\begin{cases} k_i i = \left(I_m + \left(T_{jm}^{\top} n^{-\top} I_j n^{-1} T_{jm}\right)\right) \ddot{\theta}_m + D_m \dot{\theta}_m + T_{jm}^{\top} n^{-\top} \tau_j \\ L_m \dot{i} + R_m i = V_{in} - k_\omega \dot{\theta}_m \end{cases} \tag{5.20}$$

If we consider the system dynamic, the control input that allows to generate motor torque is the input voltage to the motor spires V_{in}. Setting $V_{in} \neq 0$ cause the rise of the current i_m flowing through the motor spires which contributes to the generation of the motor torque.

If we assume that the electrical dynamic is faster than the mechanical dynamic, it is possible to neglect the transition of the current, and therefore the term $L_m \dot{i}$ of Eq. (5.20). If this assumption is verified, the dynamic of the electro-mechanical coupled system can be written as:

$$k_i \frac{V_{in} - k_\omega \dot{\theta}_m}{R_m} = \left(I_m + \left(T_{jm}^{\top} n^{-\top} I_j n^{-1} T_{jm}\right)\right) \ddot{\theta}_m + D_m \dot{\theta}_m + T_{jm}^{\top} n^{-\top} \tau_j \tag{5.21}$$

where, as previously mentioned, V_{in} is the input voltage that we can directly control to generate motor torque.

Let us consider the goal of controlling pure joint movements, in the sense that we are interested in assigning a proper control input to the motor, such that dynamic of the transmission becomes *invisible* from the point of view of high level commands. We can therefore assign to the input voltage V_{in} a control input u of the form:

$$u = \frac{R_m}{k_i} \left[\left(I_m + \left(T_{jm}^{\top} n^{-\top} I_j n^{-1} T_{jm}\right)\right) n T_{jm} y + (D_m + k_\omega) \dot{\theta}_m + T_{jm}^{\top} n^{-\top} \tau_j \right] \tag{5.22}$$

If the control input u is dimensionally consistent with the input voltage V_{in}, we can directly assign $u = V_{in}$ in order to obtain a dynamic of the controlled system of the form of:

$$\ddot{\theta}_m = nT_{mj}^{-1}y \tag{5.23}$$

which, considering Eq. (5.15) represent the decoupled dynamic of the system, where y is a new control input which allows to directly control the joint accelerations in the joint space:

$$\ddot{\theta}_j = y \tag{5.24}$$

It must be noted that, for a proper implementation on the DSPs, u and V_{in} must be scaled of an integer quantity α, which allows to convert the quantities necessary for control into machine units. The controlled system becomes:

$$u = \alpha\frac{R_m}{k_i}\left[\left(I_m + \left(T_{jm}^{\top}n^{-\top}I_j n^{-1}T_{jm}\right)\right)nT_{jm}y + (D_m + k_{\omega})\dot{\theta}_m + T_{jm}^{\top}n^{-\top}\tau_j\right] \tag{5.25}$$

In the following sections will be defined the new control input such that the system takes the desired behavior. In particular, starting from the definition of position control law, also torque control and impedance control will be defined.

5.2.1 Position Control

The control of the position can be achieved by assigning a control input y of the form:

$$y = \ddot{\theta}_j^d + K_d(\dot{\theta}_j^d - \dot{\theta}_j) + K_p(\theta_j^d - \theta_j) \tag{5.26}$$

The control of Eq. (5.26) requires the knowledge of the position error ($e = \theta_j^d - \theta_j$) and its derivative. Moreover, it requires the desired acceleration of the system $\ddot{\theta}_d$ as a feed-forward term. If substituted to Eq. (5.24), the overall controlled system dynamics takes the form of:

$$\ddot{e} + K_d\dot{e} + K_p e = 0 \tag{5.27}$$

Equation 5.27 represents the second order dynamics of the tracking error, which is asymptotically stable for $K_d > 0$ and $K_p > 0$.

If the model of the system is not perfectly known, and therefore a proper cancellation of the system dynamics cannot be accurately performed, another kind of controller is preferred. More precisely, when errors are present in the inverse dynamic cancellation of the system the feedback linearization brings to a dynamic of the form:

$$\ddot{\theta}_j = y + \zeta \tag{5.28}$$

where ζ takes into account all the disturbances due to the presence of errors in Eq. (5.25). This issue can be solved by means of exploiting more complicated control stategies that take into consideraion the disturbances introduced by the model errors in the control inputs. This is the case of the robust control shown in [6]. In this sections the problem is solved by means of adding an integral component of the position error to the control strategy.

If we assign y of the form:

$$y = \ddot{\theta}_j^d + K_d \dot{e} + K_p e + K_i \int e \tag{5.29}$$

and if we change variables such that $x = \int e$, the overall system dynamics takes the form of:

$$\dddot{x} + K_d \ddot{x} + K_p \dot{x} + K_i x = \zeta \tag{5.30}$$

which shows that at steady-state, the disturbance ζ influences the term $K_i x$, while $\dot{x}$ goes to zero. Considering finally that $\dot{x} = e$, we have obtained a null steady state error.

5.2.2 Torque Control

When instead we are interested in assigning to the controlled system a behavior that responds to externally applied forces, it is possible to define a control input y of the form:

$$y = I_d^{-1}(\tau_j^d - \tau_j) - I_d^{-1} D_d \dot{\theta}_j \tag{5.31}$$

A control input y as the one proposed in Eq. (5.31) allows, in fact, to obtain a controlled system behavior of the form:

$$I_d \ddot{\theta}_j + D_d \dot{\theta}_j = \tau_j - \tau_j^d \tag{5.32}$$

The controlled system deriving from the application of the above control input (5.31) behaves as a damped mass, given the torque error $e_\tau = \tau_j^d - \tau_j$. At steady state, the system moves with a velocity that depends on the term D_d as $\dot{\theta}_{ss} = \frac{1}{D_d} e_\tau$.

It is remarkable here that τ_d is a torque reference, which needs to be tracked. Looking in details at Eq. (5.31), the torque error $e_\tau = \tau_j^d - \tau_j$ is controlled by means of pure proportional gain $Kp = I_d^{-1}$. We can in fact here write Eq. (5.31) as:

$$y = K_p(\tau_j^d - \tau_j) - K_d \dot{\theta}_j^d \tag{5.33}$$

where $K_d = I_d^{-1} D_d$. If we are interested to control the torque error, i.e. we are interested in obtaining a null steady state torque error, the control strategy can be modified as follows:

$$y = K_p(\tau_j^d - \tau_j) + K_i \int (\tau_j - \tau_j^d) - K_d\dot{\theta}_j^d \tag{5.34}$$

where the integral term guarantees null torque error at steady state.

5.2.3 Impedance Control: Classical

A pretty classical approach to impedance control exploiting feedback linearization is presented here. Implementing impedance control means that we want to assign to the controlled system a behavior of a mass, spring and damping system which moves connected to a desired trajectory. When a force acts on this mass, the system dynamically respond to the disturbance moving to a new equilibrium position which depends on the applied force as $e = \frac{F_{ext}}{K_s}$, being F_{ext} the externally applied force, and K_s the stiffness of the desired spring.

In practice it is necessary to assign a controlled input of the form:

$$y = \ddot{\theta}_j^d + I_d^{-1}(D_d(\dot{\theta}_j^d - \dot{\theta}_j) + K_d(\theta_j^d - \theta_j) + (\tau_j - \tau_j^d)) \tag{5.35}$$

Compared to Eq. (5.26), it is noticeable the addition of a force feedback term which allows to obtain a compliant behavior to the application of external forces. By applying y, the controlled system takes the form:

$$I_d\ddot{e}_j + D_d\dot{e}_j + K_d e_j = e_\tau \tag{5.36}$$

being $e_\tau = \tau_j - \tau_j^d$. In this context τ_d is a torque reference which can be used, for example, to cancel the gravitational component of torques to the feedback control algorithm. If we consider in fact a static situation, the steady state position error of the controlled system is $e_j = \frac{e_\tau}{K_d}$. If we set $\tau_j^d = 0$, the steady state position error becomes $e_j = \frac{\tau_j^G}{K_d}$, being τ_j^G the gravitational component of torque measured by the (*virtual*-)joint torque sensors. If we set instead $\tau_j^d = \hat{\tau}_j^G$, being $\hat{\tau}_j^G$ an estimation of the gravitational component of the joint torque vector, the steady state position error becomes $e_j = \frac{\tau_j^G - \hat{\tau}_j^G}{K_d} = \frac{\zeta}{K_d}$, being ζ a small error due to the sensors noise.

Another way to achieve the same behavior which, from an implementation point of view, allows to reuse part of the code which performs torque control, is to reuse Eq. (5.33) and to assign the spring and damping behavior through the control input τ_d. We can in fact define:

$$\begin{cases} y = I_d^{-1}(\tau_j^d - \tau_j) \\ \tau_j^d = I_d\ddot{\theta}_j^d + D_d(\dot{\theta}_j^d - \dot{\theta}_j) + K_d(\theta_j^d - \theta_j) + \hat{\tau}_d \end{cases} \tag{5.37}$$

where $\hat{\tau}_d$ becomes a new torque reference which can be used, for example, to compensate for the gravitational term.

Also in this case, model errors can give origin to disturbance terms that do not allow to perfectly achive null steady state error. A solution to this problematic can be achieved by means of a small modification of Eq. (5.34) as will be presented in next section.

5.2.4 Impedance Control: The Role of Integral

Let us now consider a torque regulator of a form similar to the one presented in Eq. (5.34):

$$y = K_p(\tau_j^d - \tau_j) + K_i \int (\tau_j^d - \tau_j) - K_i K_p \dot{\theta}_j^d \tag{5.38}$$

where $K_p = I_d^{-1}$. Let us define the desired torque τ_d of the form of:

$$\tau_j^d = I_d \ddot{\theta}_j^d + D_d(\dot{\theta}_j^d - \dot{\theta}_j) + K_d(\theta_j^d - \theta_j) + \hat{\tau}_j^d \tag{5.39}$$

If we substitute this control input in Eq. (5.38) and y in Eq. (5.28), and if we define the variable $e_{\hat{\tau}} = \tau_j - \hat{\tau}_j^d$ where $\hat{\tau}_j^d$ is a new torque reference, we obtain a dynamic of the error of the form of:

$$I_d \ddot{e}_j + D_d \dot{e}_j + K_d e_j + K_i D_d \int \dot{e}_j + K_i K_d \int e_j = e_{\hat{\tau}} + \int e_{\hat{\tau}} + K_i I_d \int \ddot{\theta}_j^d - K_i K_p \dot{\theta}_j^d + \zeta \tag{5.40}$$

and if we change the variable name, such as $x = \int e_j$, the resulting dynamics takes the form of:

$$I_d \dddot{x} + D_d \ddot{x} + (K_d + K_i D_d)\dot{x} + K_i K_d x = e_{\hat{\tau}} + \int e_{\hat{\tau}} + \zeta \tag{5.41}$$

The stability of this system depends on the eigenvalues of the third order system and it is not here analyzed. It is remarkable to notice that the overall stiffness of the controlled system becomes dependent on the desired value K_d, on the integral gain of the torque regulator and on the desired damping of the impedance relationship. Moreover, the coupled dynamics may introduce oscillatory behaviors that are due to the presence of the integral of the torque error and on the integral of the position error that, due to possible errors in the model of the system, might be in contrast. Depending on the dynamic (the gains) of the controlled system, stable behaviors can achieved.

Nevertheless, it is important to notice that the presence of un-modeled effects, such as coulomb friction, might induce to cycle limits that do not allow to properly control the system position at steady state.

5.2.5 Considerations

From Eq. (5.19) and the control strategies presented in the chapter, it is possible to define different situation which are important for an easy implementation on the control boards.

- In case of high reduction ratio n of the gear boxes, the term $J_j \frac{\left(T_{mj}^{\top} T_{mj}\right)^{-1}}{n^2}$ can be neglected.
- The term $T_{jm}^{\top} n^{-\top} \tau_j$ is difficult to be modeled in the control input of Eq. (5.22). This term has not been implemented on the real robot because, due to errors in the model based approach shown within this thesis, the cancellation did not give the desired effect. To use this term, an estimator of the relationship between the input voltage u and the *virtually* measured torque τ_j would be necessary.
- It has been empirically observed (see Fig. 5.1) that the back-emf compensation contributes to the rising of the performances for both the torque control and also impedance control. When impedance controlled, the damping gain D_d give an important contribution to the tracking of position trajectories. When the stiffness is set to a low value, small errors lead to noticeable errors in the tracking of the trajectory. An high damping gain on the derivative of the position error allows to reduce the tracking error during motion tasks. Moreover, low damping is not desired because it introduces overshoots. Figure 5.1 shows the effects of the back-emf and of the damping gain D_d on the norm of the tracking error, given a desired trajectory. Figure 5.2 instead shows the trajectory tracking for different values of back-emf compensation gains and impedance damping D_d.

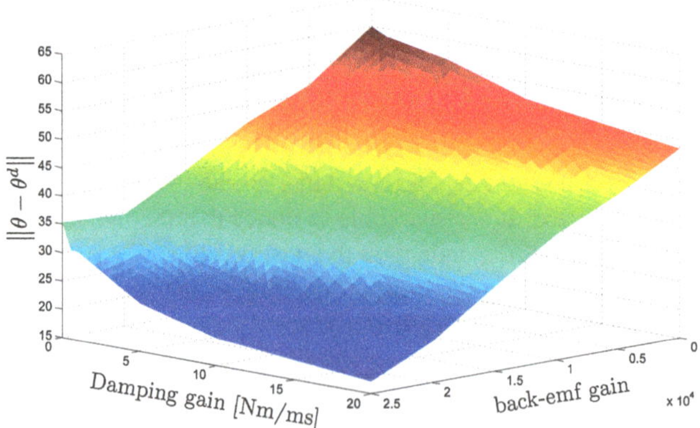

Fig. 5.1 Plot of the dependence on the back-emf and desired damping D_d, of the norm of the trajectory error $\|\theta - \theta^d\|$. The plot refers to the motion tracking of one joint controlled through the impedance relationship of Sect. 5.2.3

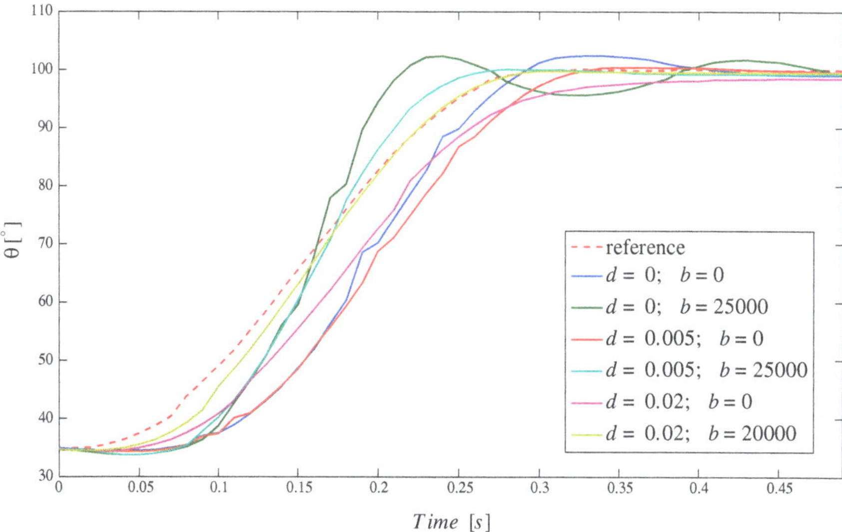

Fig. 5.2 Trajectory tracking for different values of back-emf compensation gains and impedance damping

5.3 A Case Study: The iCub Arm

In this section is presented the case of the iCub arm, which is the most significant example of coupled mechanisms present in the iCub robot [3, 5]. Other coupled mechanisms are present in the iCub wrists and torso. The wrists present a tendon driven semi-differential mechanism that perform the pitch and roll movements of the hand. The third joint, coresponding to the yaw movement of the hand, is achieved by a third motor housed in the forearm, proximal to the elbow link.

Similarly to the wrist, the torso joints movements are performed by means of a similar mechanism which emploies two motors in a differential configuration to move the roll and the pitch joints. A third motor actuates the yaw movement. Note that the adoption of differential mechanisms have the advantage of minimizing the dimensions and weight of the joints and, at the same time, obtaining an higher torque to weight ratio, despite an higher complexity in the mechanics and control.

The requisites of low weight and small dimension of the platform lead to the choice of a shoulder mechanism designed as a three DoF differential mechanism, where three motors housed in the upper torso cooperate together to actuate a universal 3DoF mechanism. Therefore, while performing the motion of the shoulders the motors' housings are not moved, resulting in a reduced moving inertia characterizing the arm structure.

The shoulder joint is a cable differential mechanism with a coupled transmission system (see Fig. 5.5). Three parallel motors housed in the upper-torso move pulleys to generate the spherical motion of the shoulder (see [4, 7] for a more clarifying

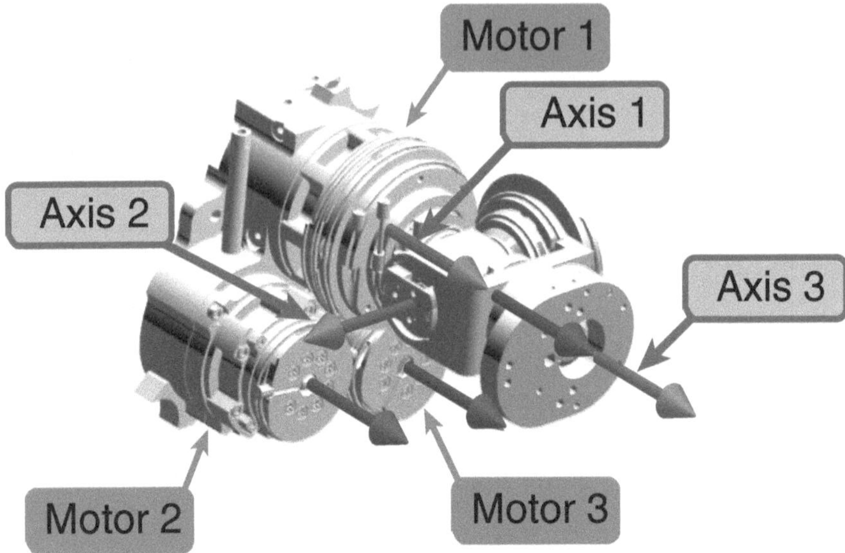

Fig. 5.3 Particular of the iCub shoulder. A CAD view of the shoulder joint mechanism showing the three motors actuating the joint and the pulley system

explanation). Figure 5.4 shows a bi-dimensional schematic representation of the transmission of the motion from the motors to the joints.Absolute hall effect based sensors measure the angular position of the joints, after the reduction gearboxes. The resolution of these sensors is 4092 tics per round. Motor position are measured instead with the hall effect sensors that are placed internally, in the spirals of the brushless motor. These sensors are used to commutate the current flowing through the spires of the motor. Three hall effect sensors measure 8 changes in the magnetic field of the rotor while moving, corresponding to 48 ticks (on/off signals) per round. This value should be multiplied for the reduction ratio of the gearbox, to make a comparison with the joint encoder resolution (the reduction ratio is 1:100, for a total of 4800 tics per round). Motor encoders are not only used for performing the commutation of the phases, but also to perform the compensation of the *back-emf* torque of the motosr.

The motor group is constituted by brushless frameless motors (RBE Kollmorgen series) with harmonic drive reductions (CSD series with 100:1 ratio) [7] and are located in the torso frame. A first bigger actuator (Motor 1 in Fig. 5.5) is capable of delivering 40 Nm and two medium power motors (Motor 2 and Motor 3) provide 20 Nm each. The motor group is constituted by brushless frameless motors (RBE Kollmorgen series) with harmonic drive reductions (CSD series with 100:1 ratio) [7] and are located in the torso frame. A first bigger actuator (Motor 1 in Fig. 5.5) is capable of delivering 40 Nm and two medium power motors (Motor 2 and Motor 3) provide 20 Nm each. The first motor actuates directly the first joint (shoulder pitch),

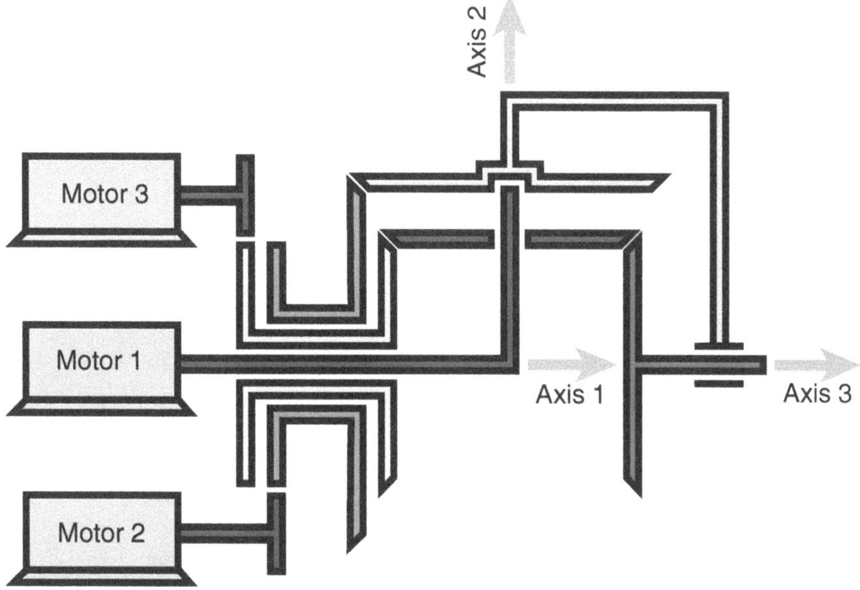

Fig. 5.4 Sketch of the working principle of the mechanism of Fig. 5.3

whereas the second and third motors actuate two pulleys that are coaxial with the first motor.

The elbow joint has an independent frameless brushless motor. The joint is commanded with tendons in push-pull configuration, moving an idle pulley. The position of the joints is controlled by means of an absolute hall effect based encoder placed on the joint side, while hall effect sensors placed inside the motor spires are employed to perform the commutation and to compensate for the *back-emf* component.

The wrist is a 3DoF mechanism. The roll movement (i.e. the movement competing for the pronosupination of the wrist) is achieved by means of a single brushed motor directly connected to the forearm. The pitch and yaw movements instead are accomplished by means of two motors which move a semi-differential tendon driven mechanism.

The iCub shoulder kinematic coupling matrix is constant and depends on the ratio among the radius of the pulleys moved by each motor. This operator takes the form of (5.42)

$$T_{mj} = \begin{bmatrix} 1 & 0 & 0 \\ 1 & a & 0 \\ 0 & -a & a \end{bmatrix}, \tag{5.42}$$

where a is a constant value which depends on the dimension of the pulleys. For the iCub shoulder $a = 40/65 \approx 0.6154$. The inverse relationship is:

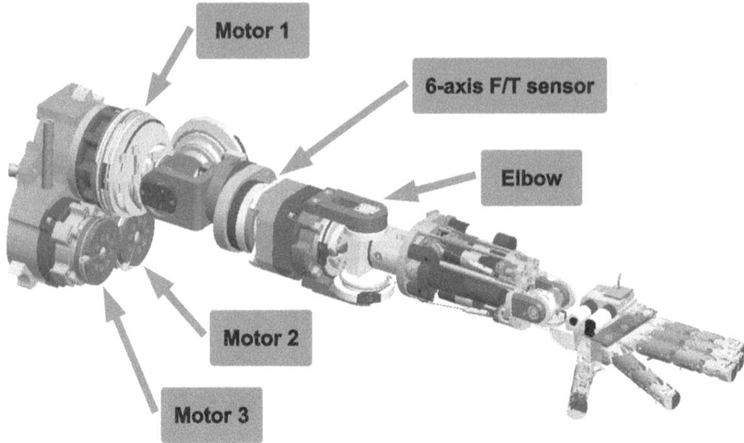

Fig. 5.5 The iCub arm. A CAD view of the shoulder joint mechanism showing the three motors actuating the joint and the pulley system, the F/T sensor, the elbow and the hand

Table 5.1 Datasheet parameters of the motors actuating the shoulder mechanism

	M_1	M_2	M_3
$R_m[\Omega]$	0.664	0.698	0.698
$k_i[\frac{Nm}{Amp}]$	0.0410	0.0236	0.0236
$k_\omega[\frac{Amp\cdot s}{rad}]$	0.0410	0.0236	0.0236
$I_m[Kgm^2]$	$9.2e-6$	$9.2e-6$	$9.2e-6$
$D_m[\frac{Nms}{rad}]$	$2.09e-6$	$9.18e-7$	$9.18e-7$
N	100	100	100
$h[\frac{Duty}{V}]$	33.25	33.25	33.25
$\gamma[\frac{tics}{rad}]$	651.9	651.9	651.9

$$T_{mj} = \begin{bmatrix} 1 & 0 & 0 \\ -r & r & 0 \\ -r & r & r \end{bmatrix}, \tag{5.43}$$

being $r = 1/a = 1.625$.

The datasheet parameters of the shoulder's motors are reported in Table 5.1.

In order to implement the inverse dynamic on the coupled mechanism (see Sect. 5.2), let us define a control input of the form:

$$u = I_U y + D_U \Theta_M + T_\tau \tau_j \tag{5.44}$$

being

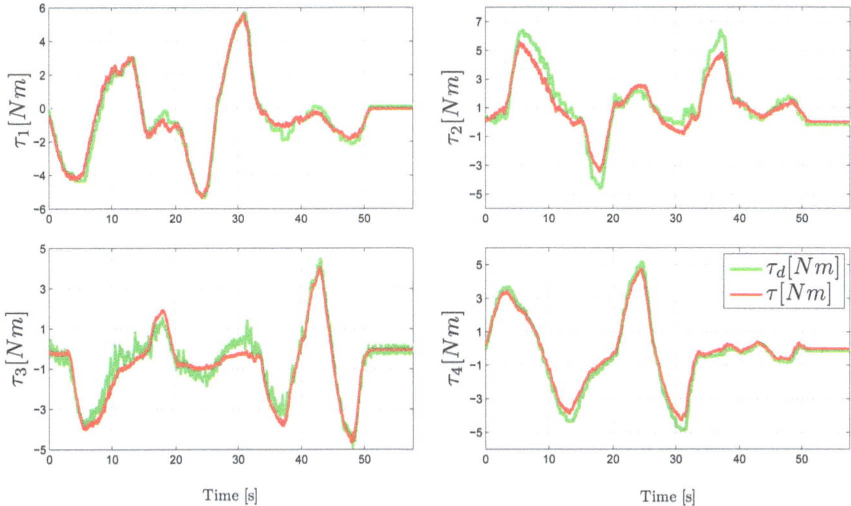

Fig. 5.6 Torque control: desired (*green line*) versus actual (*red line*) external joint torques. It is here shown the torque regulation resulting from the application of the controller proposed in Sect. 5.2.3. The method has been applied to four joints of the iCub right arm. The desired torque τ_d derives from the impedance regulator as described in Eq. (5.37)

$$I_U = \begin{bmatrix} \alpha \frac{NR_1 I_{m1}}{k_{i1}} & 0 & 0 \\ -\alpha \frac{NR_2 I_{m2}}{k_{i2} a} & \alpha \frac{NR_2 I_{m2}}{k_{i2} a} & 0 \\ -\alpha \frac{NR_3 I_{m3}}{k_{i3} a} & \alpha \frac{NR_3 I_{m3}}{k_{i3} a} & \alpha \frac{NR_3 I_{m3}}{k_{i3} a} \end{bmatrix} \tag{5.45}$$

$$D_U = \begin{bmatrix} \frac{1}{\gamma} \frac{R_1(D_{m1}+k_{\omega 1})}{k_{i1}} & 0 & 0 \\ 0 & \frac{1}{\gamma} \frac{R_2(D_{m2}+k_{\omega 2})}{k_{i2}} & 0 \\ 0 & 0 & \frac{1}{\gamma} \frac{R_3(D_{m3}+k_{\omega 3})}{k_{i3}} \end{bmatrix} \tag{5.46}$$

After this compensation, the system is capable of behaving as preferred, by assigning to y the proper control strategy. It can be a stiff system, as proposed in Sect. 5.2.1. It can behave as a damped mass, as shown in Sect. 5.2.2. Or it can simulate a mass, spring and damper system, as in the case of Sect. 5.2.3.

The control strategy used to present some of the results obtained with the iCub robot is the one of Sect. 5.2.3, shown in Eq. (5.37). This control strategy implements both torque control, and assigns a reference torque such as the overall behavior of the system is the one of a mass, spring and damper mechanism. Figure 5.6 shows the torque tracking while the robot is interacting with the environment.

An external force is applied at the end-effector of the robot and is measured by means of the FTSs embedded in the third link of the arm, just after the shoulder. The torques are measured through the algorithm of Chaps. 3 and 4. These *virtual joint torque measurements* [1, 2, 5] are used as feedback to perform the control.

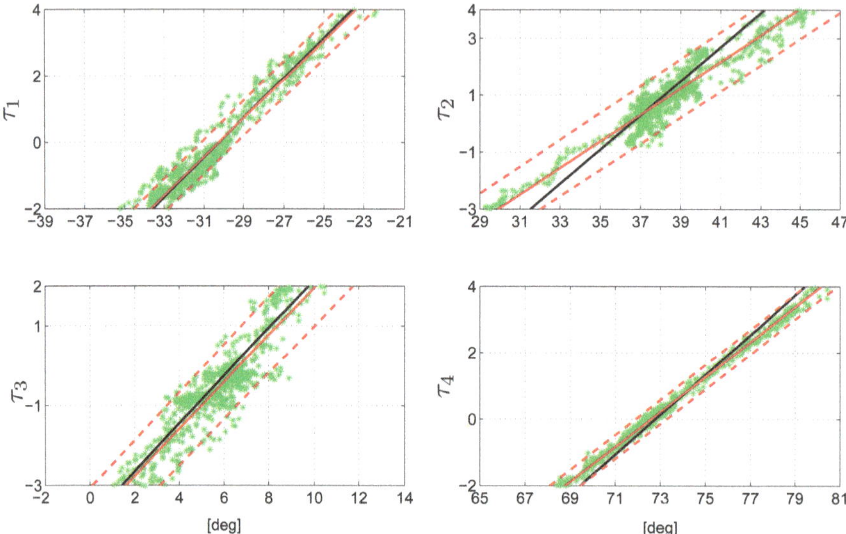

Fig. 5.7 Impedance control: desired (*black solid line*) and measured (*red solid line*) stiffness resulting from the application of the impedance controller Eq. (5.37) to four different joints of the iCub right arm in $q_d = [-30, 37, 6, 73]^\top$. The measured line is the result of linear fitting the measured data points (*represented by green dots*). A 95 % confidence interval for the measured stiffness is represented with *red dashed lines*

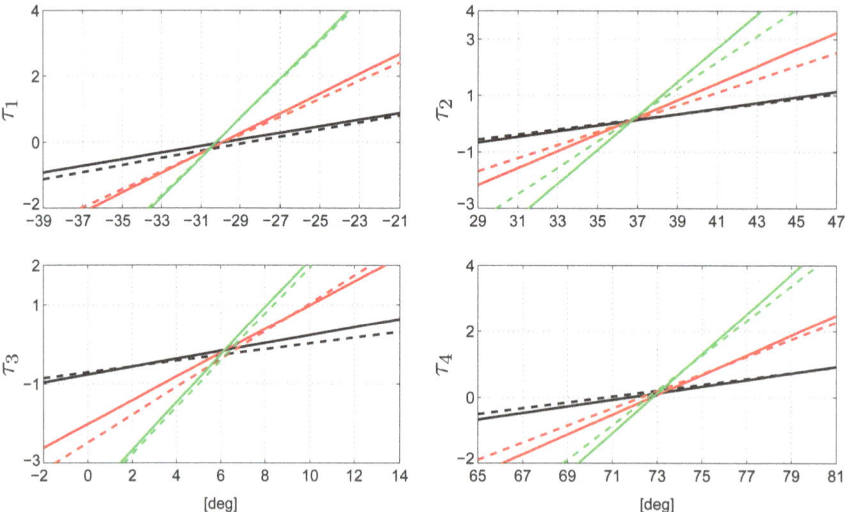

Fig. 5.8 Impedance control: desired (*solid lines*) and measured (*dashed lines*) stiffness resulting from the application of the impedance controller Eq. (5.37) on four joints of the right arm in $q_d = [-30, 37, 6, 73]^\top$. Three different stiffness have been simulated: $k_d = 0.1 \frac{Nm}{rad}$ (*black lines*), $k_d = 0.3 \frac{Nm}{rad}$ (*red lines*), $k_d = 0.6 \frac{Nm}{rad}$ (*green lines*)

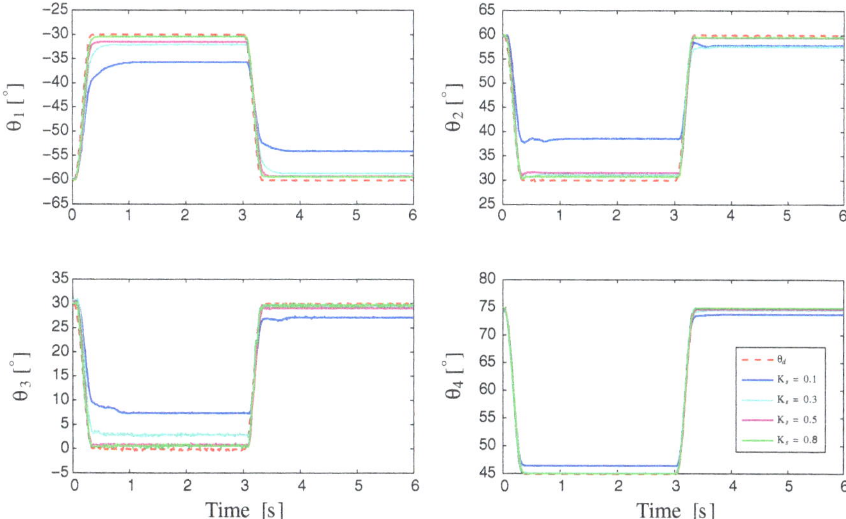

Fig. 5.9 Comparison of the trajectory tracking with integral gain of the torque controller $K_i = 0$, for different values of desired stiffness K_s

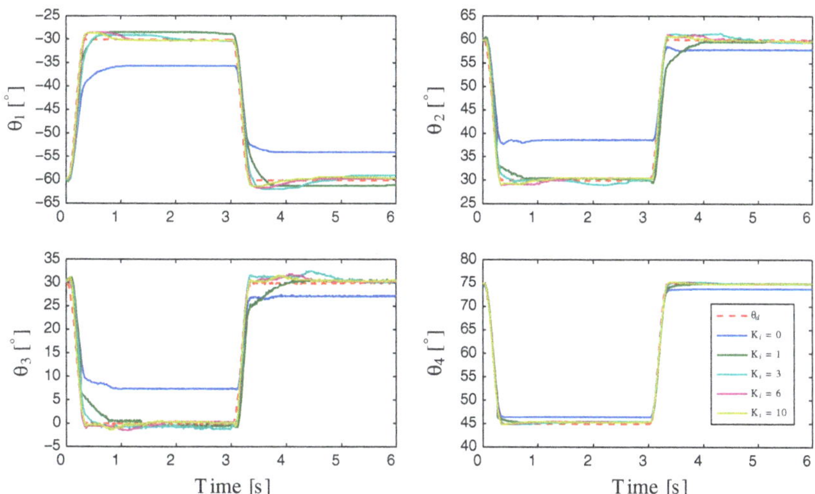

Fig. 5.10 Comparison of the trajectory tracking with desired stiffness $K_s = 0.1$, with different values of the integral gain of the torque controller

The joint level control is performed on the motor control boards at the rate of 1kHz. Chapter 6 will show how these information are computed, and the path the information follow in order to obtain a measurement to perform feedback control on the motor control boards.

Figure 5.6 shows the commanded torques, which depend on the desired impedance behavior of the system. In particular, they refer to the stiffness control reported in Fig. 5.7. Referring to Eq. (5.36) a desired K_d have been set for the shoulder and elbow joints equal to $K_p = 0.6[\frac{Nm}{rad}]$. Rearranging the points of Fig. 5.6 together with encoder data, the validation of the method has been performed.

In Fig. 5.8, the experiment have been repeated assigning different values to the desired stiffness K_d. It is remarkable the possibility to vary this parameter through high level software commands.

Experiments refering to the performances during trajectory tracking in impedance control mode are shown in Fig. 5.9, for different values of desired stiffness, and in Fig. 5.10, where the role of the integral component on the torque feedback is highlighted.

References

1. Fumagalli, M., Ivaldi, S., Randazzo, M., Natale, L., Metta, G., Sandini, G., et al. (2012). Force feedback exploiting tactile and proximal force/torque sensing—Theory and implementation on the humanoid robot iCub. *Autonomous Robots*, *33*, 381–398.
2. Ivaldi, S., Fumagalli, M., Randazzo, M., Nori, F., Metta, G., & Sandini, G. (2011). *Computing robot internal/external wrenches by means of inertial, tactile and F/T sensors: Theory and implementation on the iCub*. 2011 11th IEEE-RAS International Conference on Humanoid Robots (Humanoids), 521–528.
3. Natale, L., Nori, F., Metta, G., Fumagalli, M., Ivaldi, S., Pattacini, U., et al. (2012). The iCub platform: a tool for studying intrinsically motivated learning. Baldassarre, Gianluca and Mirolli, Marco (Eds.), *Intrinsically motivated learning in natural and artificial systems*. Berlin: Springer.
4. Parmiggiani, A., Randazzo, M., Natale, L., Metta, G., & Sandini, G. (2009). *Joint torque sensing for the upper-body of the iCub humanoid robot*. International Conference on Humanoid Robots, Paris, France.
5. Randazzo, M., Fumagalli, M., Nori, F., Natale, L., Metta, G., & Sandini, G. (2011). *A comparison between joint level torque sensing and proximal F/T sensor torque estimation: Implementation on the iCub*. 2011 IEEE/RSJ International Conference on Humanoid Robots (Humanoids), Intelligent Robots and Systems (IROS), 4161–4167.
6. Sciavicco, L., & Siciliano, B. (2005). *Modelling and Control of Robot Manipulators.Avanced Textbooks in control and Signal Processing*. Berlin: Springer.
7. Tsagarakis, N., Metta, G., Sandini, G., Vernon, D., Beira, R., Santos-Victor, J., et al. (2007). icub—the design and realization of an open humanoid platform for cognitive and neuroscience research. *International Journal of Advanced Robotics*, *21*(10), 1151–1175.

Chapter 6
Hardware and Software Architecture

Abstract The chapter presents an overview of the hardware and software architecture of the iCub robot. In particular, it shows the way the code to perform torque feedback and active compliance control have been implemented. Qualitative peformances of the implemented methodologies are reported by means of a comparison of the *virtal sensor* method, with mechanical joint torque sensors and additional external FTS. Based on the YARP famework, a set of modules perform the estimation of joint torques, of external forces and of gravity compensation terms have been implemented. The implementation of these YARP modules are finally reported.

6.1 YARP

YARP (Yet Another Robot Platform) is a set of open source, OS-independent libraries that supports distributed computation under different operative systems (Windows, Linux, Mac) with the main goal of achieving efficient robot control [9].

One of its main feature is that it also supports hardware and software modularity. Hardware modularity is obtained by defining interfaces for classes of devices in order to wrap native code API. In this way, a change in hardware requires only a change in the API calls with the advantage of decoupling the programs from the specific hardware (using Device Drivers) and operative system (relying on the OS wrapper given by ACE [13, 14]). Software modularity is achieved by providing an interprocess communication protocol based on ports, which allows the user to subdivide the main task of the robot in simple, reusable modules each of them providing specific functionalities (e.g. object tracking, grasping etc.) [5]. The user application is then obtained by interconnecting at run-time these software modules, generally running on different machines on a common network.

YARP have been used to implement the method of the virtual force sensor on he iCub robot and the compliant control system as a collection of interconnected independent modules. The reasons to define such complex architecture for estimation and

M. Fumagalli, *Increasing Perceptual Skills of Robots Through Proximal Force/Torque Sensors*, Springer Theses, DOI: 10.1007/978-3-319-01122-6_6,

control are practical: firstly, one single CPU, although powerful, can never be enough to cope with more and more demanding applications; secondly, smaller subsystems and executables are easier to be maintained and updated, and their employment makes the overall system handy.

6.2 iCub

This section presents the electronic hardware components of the iCub robot, partly already introduced in Chap. 2.

6.2.1 The iCub Hardware Architecture

A cluster of standard PCs and a Blade system are interconnected through a 1GB ethernet and constitute the core of the brain of iCub. A server connects these computers and user PCs over a common network. These machines are generally dedicated to the high-level software computation which is more demanding (e.g. coordinated control, visual processing, learning, cartesian interfaces). Low level software interfaces, data sharing and sensors acquisition is performed on a dedicated computer mounted on the head of the iCub robot, a *pc104*. Motor control is implemented on the DSPs embedded on the boards which are present on the robot body.[1] Table 6.1 shows the main hardware components constituting the iCub *sensori-motor* system[2] [10].

Motors and encoders are connected to the motor control boards used for encoders data acquisition and precise joint level motion control.[3] Motor control boards and force/torque sensors are connected through CAN-bus lines. A total of seven CAN-Bus lines allow the communication between the motor control boards, force sensors, and a central control unit (see Fig. 6.1). This network of CAN lines converges to an electronic board, namely the *cfw2* which manages the flow of information between the sensori-motor system of the robot, and its central control unit, the pc104. The *cfw2* board is characterized by a set of 8 available CAN bus connections, 2 firewire ports and 2 microphone ports. It implements a fast CAN communication (full send and receive bandwidth: 6–8 messages/1 ms). The pc104 have a complete description of the entire *sensori-motor* system provided by sanity check messages generated by the DSP boards, which broadcast a message describing their current state (sensors status, external faults, communication failures, overloads etc.) every fixed amount of time

[1] Low level code runs on Freescale DSP56F807 which is built with Freescale CodeWarrior Development Studio, which is a complete integrated Development Environment (IDE) that provides a highly visual and automated framework to accelerate the development of embedded applications [2].

[2] Table 6.1 shows only the main hardware components (motors and sensors) which have been used within this work, and the main components of the of the robot perceptive system.

[3] The coupling of the motors of the iCub shoulders, shown in Sect. 5.3, is performed on the DSPs.

Table 6.1 The hardware components of the iCub *sensori-motor* system

Sensor/Actuator	Brand	Model	Position	Main specifications
Brushless DC motor	Kollmorgen	RBE-01210-A frameless	Shoulder, elbow, legs, torso	$Tc = 0.115$ Nm, $Ic = 5.41$ A
Motor control board	Custom IIT	BLL–BLP	Torso, legs	2 DC brushless motor control boards
Brushed DC motor	Faulhaber	DC micromotors 12xx G series	Hands, head	Motor, gear box, encoder; 0.2–0.6 mNm
Motor control board	Custom IIT	MC4–MCP	Torso, arms, head	4 DC brushed
Absolute encoder	Honeywell	SS495A1	Joint axis	Hall effect sensor, 8.7 mA, 4.5–10.5 Vcc, p.1.3 mm
3 DOF orientation tracker	Xsens	MTx-Miniature 3D inertial tracker	Head	12bits, 1.7–5 g,
6-axis force/torque sensor	Custom IIT	–	Arms and legs	16 bit, microchip 16Fxx microcontroller,16 bit A/D converter @1 kHz
Cameras	Point gray	DR2-03S2C-EX-CS—Videocam dragonfly2 color extended version	Eyes	1640×480, 1/3, CCD

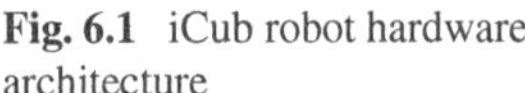

Fig. 6.1 iCub robot hardware architecture

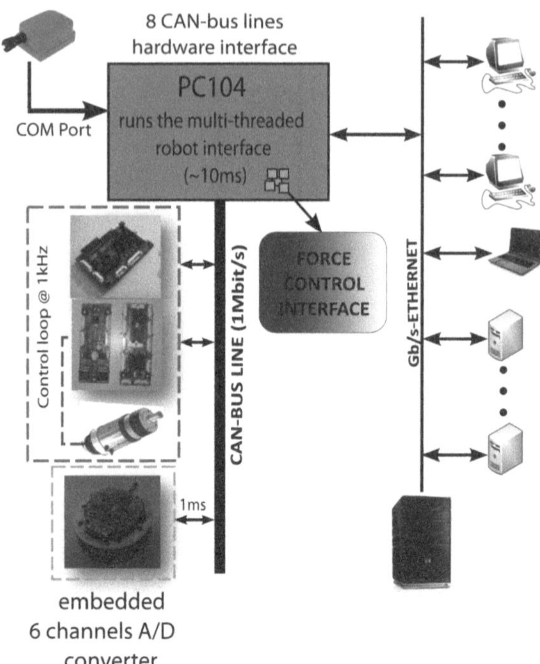

(currently 5 s). Therefore, the main control unit have a complete description of the entire *sensori-motor* system. The *cfw2* boards also provides ports for the acquisition of camera data and for the 3DOF orientation tracker by means of Firewire ports.

A PCI interface allows the communication between the *cfw2* and the *PC104*. The *PC104* board mounts the central control unit of the robot on which runs the low level software interface to the *cfw2* and to the data flowing through the CAN networks and the Firewire ports. The YARP interfaces for the devices are here used to perform hardware independent communication with the boards. Finally, the pc104 is connected to a network of cluster and computers to allow the communication between the robot and the user, where an Intel 1630 performs server operation.

6.2.2 The iCub Software Architecture

The iCub software architecture comprises a set of YARP executables which are called *YARP Modules*, or just *modules*. Each module runs on a PC, a blade or the pc104. The iCub software architecture is a repertoire of modules which communicate by means of YARP ports, to constitute the hardware infrastructure of the robot.

The iCub robot is constituted with a repertoire of very basic motion and sensing capabilities (encoders, 3DOF Orientation Tracker, FTSs, cameras, motors). A module called *iCubInterface* (refer to Fig. 6.2) maps the information of each actuator

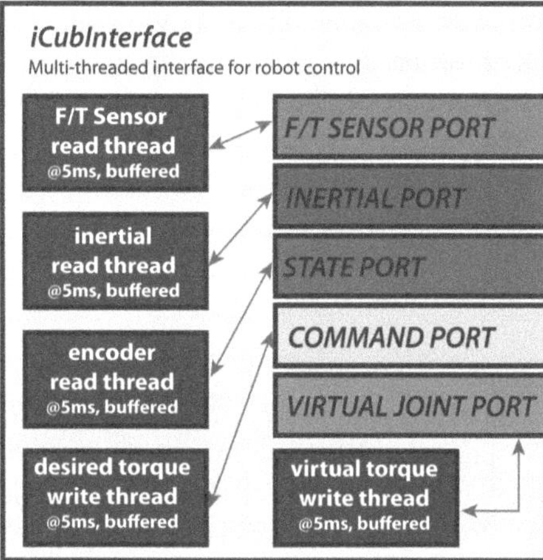

Fig. 6.2 Software interface, namely *iCubInterface* that allows the communication with low level devices (API) and allows the communication with the user through interfaces

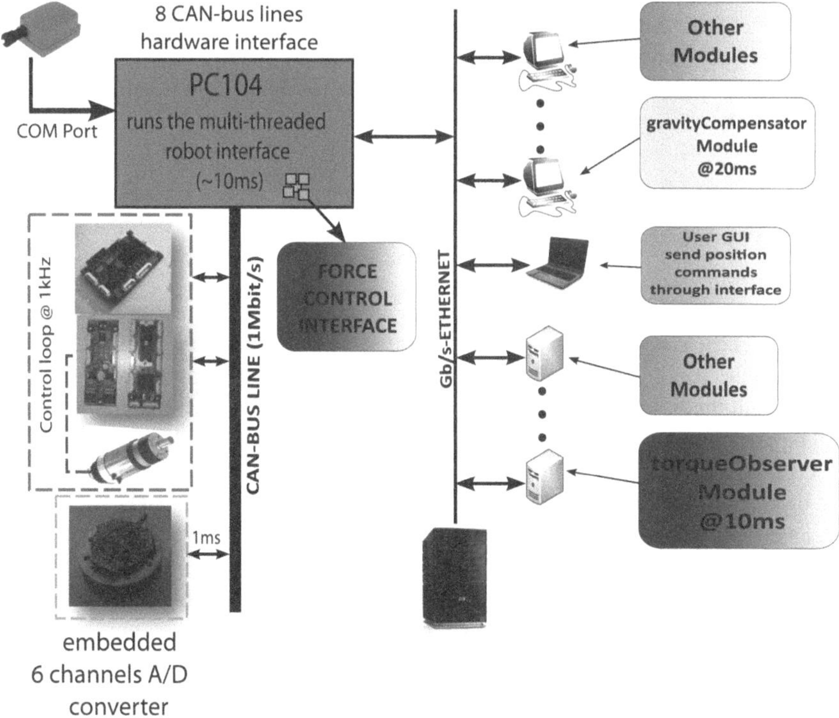

Fig. 6.3 Hardware architecture with specification of the modules and devices involved for the implementation of *virtual joint torque measurements* and control

and sensor with the corresponding joint and part of the robot. Additionally, the *iCubInterface* manages the information which flow through the *cfw2* board and opens ports to share these information with the user and other modules. It allows the user to monitor the state of the boards, to send commands and to read these information. Through these ports, the user can communicate with the robot.

6.2.2.1 iCub Modules

The software architecture of the iCub robot is a collection of YARP modules (Fig. 6.3). A module is an executable that performs a specific task and whose interface is defined by YARP ports. YARP allows to connect networks of modules, which share information and commands which constitute the behaviors of the robot. The purpose of developing software architectures is to create a core software infrastructure that enables the robot to exhibit a set of target behaviours for an Experimental Investigation (see [8]). Next sections show the methodology adopted that practically allowed to enlarge the sensorial system of the iCub robot. A set of modules have been implemented to perform contact detection, force control and gravity compensation. By means of these modules it was possible to enlarge the perception of the interaction forces, to obtain *virtual joint torque measurements* and, by means of a low level implementation of appropriate control strategies, to obtain compliant behaviors of the platform.

6.3 Force Control

Proximal F/T sensors are mounted on the iCub robot's limbs, one for each arm and leg [10]. These sensors give important information about the interaction and enlarge the tasks the robot can perform. In particular FTSs improve the perception of the robot about the outer world. It has been shown in Chap. 3 and 4 how proximal FTSs can be exploited together with the information of an artificial skin (see [1]), to have a precise, complete and distributed information about the interaction occurring during the tasks the robot performs [4, 7, 12]. From these information it is also possible to retrieve torque measurements. Torque measurements at joint level are very important for robots. They allow to implement active compliance, which is of fundamental importance for a humanoid robot interacting with unstructured environments. Moreover, a humanoid robot is supposed to perceive its surrounding environment in terms of force and touch perception, necessary for the exploration and learning of the surrounding world, and should be compliant while exploring. In this framework it is therefore necessary that the robot regulates the interaction torques at joint level.

The iCub software framework has been used to create a set of software modules which enlarge the perceptual capabilities of the robot, in terms of touch and force perceptual capabilities. The modules that have been developed and implemented allow to obtain:

- contact detection: the robot should detect critical situation of contact with the environment.
- virtual joint torque measurements: joint torque measurements allow the implementation of joint compliance and torque regulation.
- gravity compensation: necessary to improve the performances of motion and reaching tasks in terms of trajectory tracking and steady state position error.

these modules are described in next sections in terms of implementation issues, assumptions, reliability and performances (timing and delays).

6.3.1 Contact Detection

Contact detection is the basic perceptual behavior the robot is supposed to have. Basic contact detection is achieved by defining a threshold on the robot model error which assures that the robot is not interacting with the environment. Different algorithms have been developed in literature to achieve this kind of task. Most of them are complex algorithms which have the goal to reduce the interval over which the probability of a contact is high.

The iCub robot has been provided with a basic contact detection algorithm which is based on a threshold which represents a confidential interval, determined by a statistics of the difference between the model and the real measurements of the FTSs, such that:

$$\begin{cases} \text{if } \|w_{measured} - w_{model}\| > \eta \Rightarrow \xi = 1 \\ \text{if } \|w_{measured} - w_{model}\| \leq \eta \Rightarrow \xi = 0 \end{cases} \tag{6.1}$$

being ξ a boolean number defining whether a contact occurs ($\xi = 1$) or not ($\xi = 0$). As an example, Fig. 6.4 shows the detection of a table placed in front of the robot, while the end-effector moves along the z component (i.e. while performing a vertical movement). In this case, only the force vector $F \in \mathbb{R}^3$ of the wrench w has been considered, to evaluate the contact detection. When the norm of the force vector $\|F\|$ passes the threshold, the algorithm gives 1, 0 otherwise.

6.3.2 Providing Virtual Torque Measurements

The iCub robot mounts 6-axis FTSs on each arm and leg. Their position is proximal with respect to the position of the end-effector. The FTSs which are mounted on the arm are placed immediately after the $3DOF$ spherical joint of the shoulder. On the legs, they are placed close to the $2DOF$ revolutionary joint of the hip.

Preliminary versions of the iCub robot lack of joint level torque sensors. Therefore *virtual torque measurement* should be provided to perform joint level interaction control.

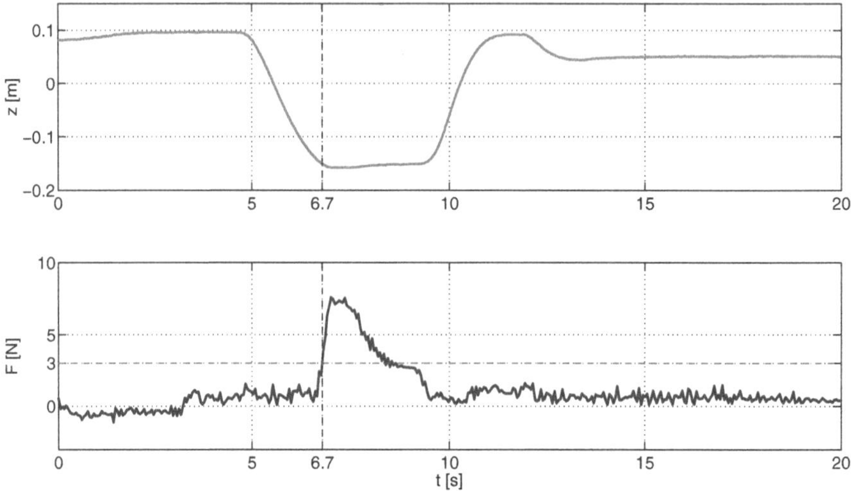

Fig. 6.4 A dataset showing the detection of a table placed in front of the iCub robot during a demonstration of the CHRIS European Project (FP7-IST-215805)

We define here *virtual measurement*, a data whose source of information is not directly extracted from a sensor measurement (a joint torque sensor, in this case) but it is computed by means of software estimation. The *virtual measurements* are directly related to one or more measurements which are performed by real sensors present on the robot.

Virtual joint torque measurements follow this principle. In Chap. 3 the formalism and the assumptions which are at the base of this method have been presented. In short, starting from the measurements of a $3DOF$ orientation tracker and joint encoders, it is possible to obtain the *virtual measurements* of links frame velocity and acceleration, given the links kinematics. On the other hand, exploiting the measurements of 6-axis F/T sensors and with an estimation of the links dynamical parameters (e.g. using a CAD model), we can address the problem of the estimation of the joint torques, by exploiting a model based approach based on enhanced oriented graphs (see Sect. 3.3).

The computation of the joint torques of the iCub robot, following the *virtual torque measurements* approach, requires a lot of effort in terms of CPU demand. These calculation cannot be performed on the DSPs of the motor control boards, characterized by limited performances and computational capabilities. For this reason, demanding computations have to be performed on the blade servers which are physically connected to the robot with a 1 GB Ethernet network. Data sharing and connection to the *iCubInterface* is achieved through YARP ports.

The *iCubInterface* module runs a thread which interacts with the low level CAN API. This thread generates the CAN messages to send to the boards using the CAN-Network. The messages contain both the ID of the board which sends the messages and the ID of the board which will receive it. It is therefore only required that the

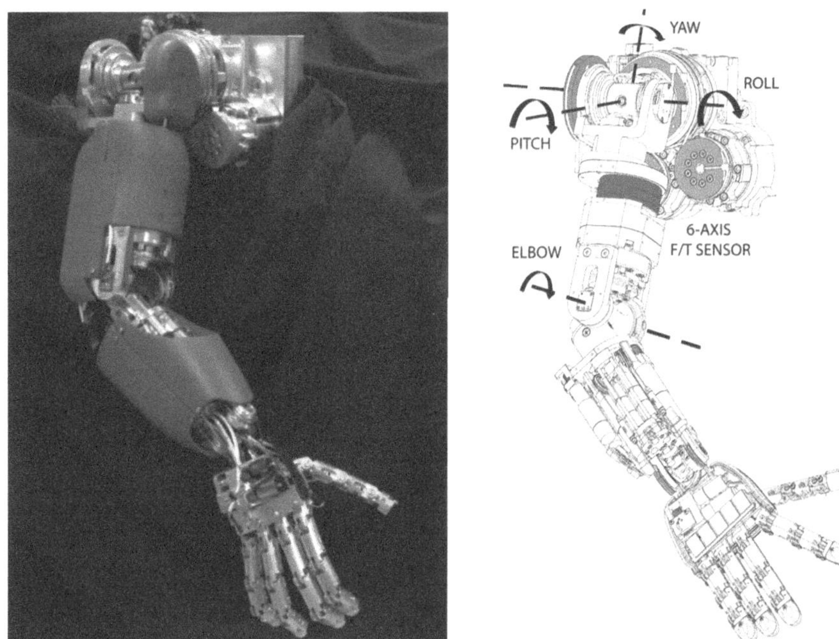

Fig. 6.5 The iCub arm of version $v2.0$. The mechanism mounts joint level torque sensors on the first four joints

motor control boards know the ID of the measurement they have to exploit, that the approach becomes independent on the sensor employed to perform the measurement. In other word, if a joint torque sensor is present, a data acquisition board will send through the can bus a message with a proper ID. Otherwise, a module running on an external machine will send the *virtual measurement* to the *iCubInterface*, which will convert this message to a CAN bus message with proper ID.

By exploiting this approach it becomes sufficient that the motor control boards which receive the messages have knowledge of the ID of the board which sends the message they are waiting for, in order to have the torque data. In this way, the method becomes transparent at the sensor level, from the point of view of the boards.

Comparison Between Torques

It is shown hereafter a comparison between the different possibilities of calculating joint torques [12]. The setup used to make this sort of study was the 4 DOF iCub shoulder version 2.0 (Fig. 6.5), which also mounts joint torque sensors on first and second joint of the shoulder (the pitch and roll) and on the elbow joint. For the measurement of the yaw joint instead, it has been decided to exploit the measurements of the 6-axis force/torque sensor, which is present also on the versions $v1.x$ of the

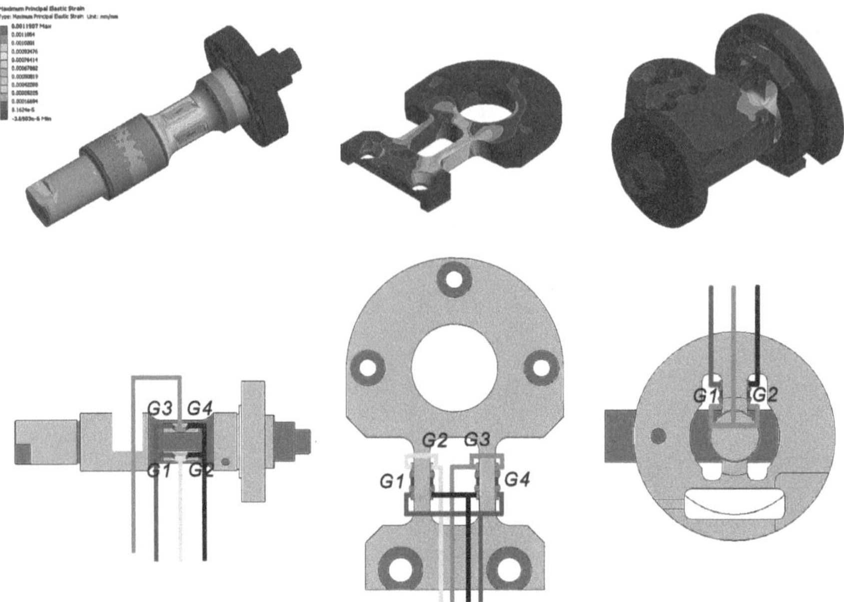

Fig. 6.6 Three joint torque sensors of the iCub shoulder $v2, 0$. From the *left*, the sensors for joint 0 (shoulder pitch), 1 (shoulder roll) and 3 (elbow). The torque sensor measurement for joint 2 (shoulder yaw) is provided by the 6-axis force/torque sensor

robot. Figure 6.6 shows the finite element analysis and the position where the strain gauges have been glued for the three custom joint torque sensors, developed in [11].

In order to evaluate the estimation of the virtual joint torques, and to perform a comparison between the virtual joint torque measurements and the real sensors, and additional $6 - axis$ FTS have been mounted at the terminal position of the forearm, as shown in Fig. 6.7.

The first experiment that is here proposed is a comparison between the torques that are due to an externally applied wrench. Given the set of sensors presented previously for the setup which is taken into consideration here, we compare the measured joint torques τ_m, using three approaches:

- $\tau_m = \tau_e = J^\top w_e$ which make use of the external force sensor, being w_e the measurement of the externally applied wrench.
- $\tau_m = \tau_v$, being τ_v the virtual measurement which exploits the method presented in previous chapters and the measurements of the proximal force/torque sensor.
- $\tau_m = \tau_s$, being τ_s the measurement of the joint torque sensors.

Figure 6.8 shows the results of an experiment where an external wrench is applied at the end-effector of the iCub arm $v2.x$ and whose value is measured by the external force/torque sensor, by the joint level torque sensors and by the *virtual joint torque sensors*. It can be noted that, a part from the offset given by the gravitational compo-

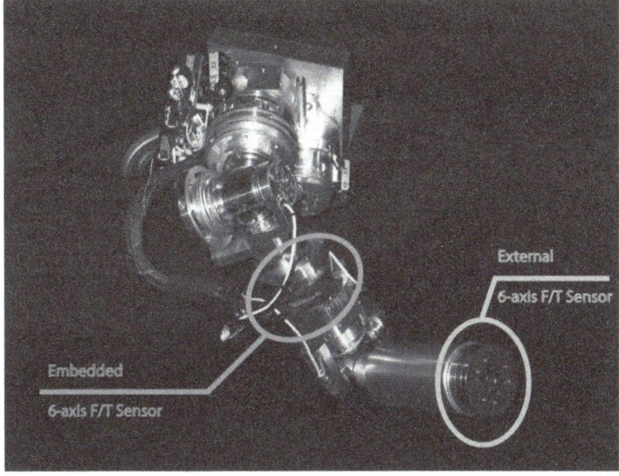

Fig. 6.7 The two 6-axis force/torque sensor employed on the *v*2.0 arm to compare external values of force and torques

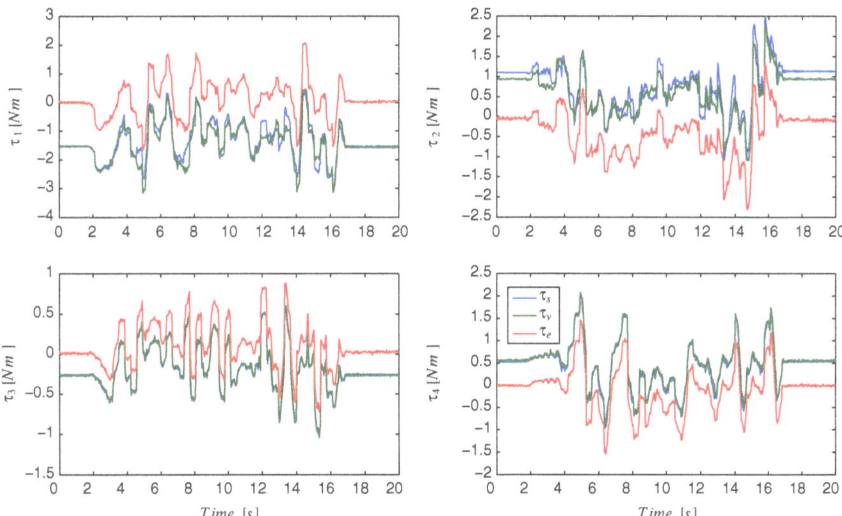

Fig. 6.8 Comparison between joint level torques. Note that the torque estimation τ_e is not affected by the initial offset given by the gravitational component of the joint torque

nent of the joint torques, of which both the virtual measurements and the joint torque sensors measurements are affected, the method of the virtual joint torque sensors gives an estimation of the joint torque that is comparable to the joint torque sensors measurements and to the projection of the external force at the joint level.

Representation of Externally Applied Wrenches

A key feature of the method shown in Chap. 3 and 4 is that it is possible to have an estimation of both the joint torques and of the externally applied wrenches. If we consider in fact the same experiment of Fig. 6.8, in which we measured the external wrenches by means of an additional FTS placed a the end-effector of the robot arm, it is possible to compare the estimation of externally applied wrenches at the end-effector, which correspond to the actual measurement of the additional FTS. These measurements are compared with both the computation of the *virtual wrench* [4, 7, 12], given the measurements of the internal force/torque sensor (i.e. the sensor mounted on the iCub shoulder), and with the inverse of the relationship $\tau = J^\top w$, being τ the measurements of joint torque sensors.

Figure 6.9 shows the result of this comparison. It is noticeable that it is not possible to have an estimation of the externally applied wrenches $w_e \in \mathbb{R}^6$ using the only measurements of joint torques $\tau_s \in \mathbb{R}^4$. The errors in the representation of the external wrench when using the joint torque sensors derive from the solution of the problem:

$$\hat{w} \; s.t. \; argmin_w \left\| J^\top w - \tau \right\| \tag{6.2}$$

whose solution is given by:

$$\hat{w} = (J^\top)^\dagger \tau + (I - (J^\top)^\dagger (J^\top))\zeta \tag{6.3}$$

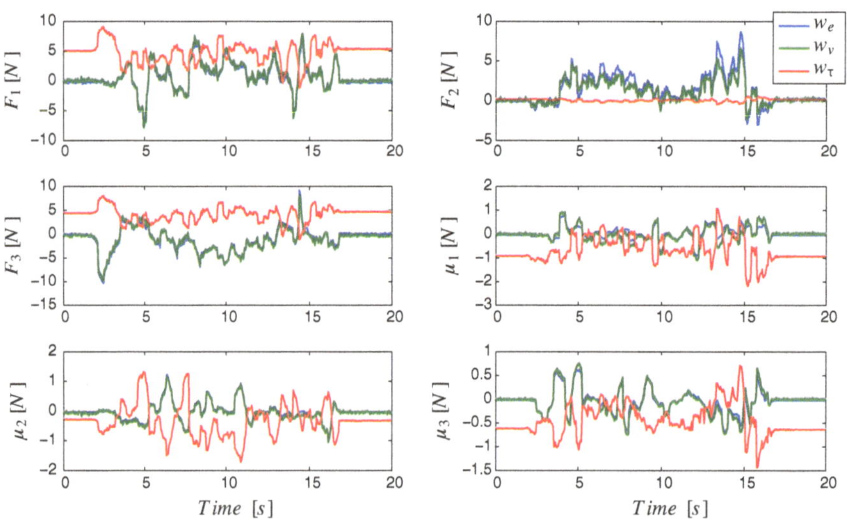

Fig. 6.9 Comparison between externally applied wrench representation. Note that the wrench estimation w_τ cannot be performed for the inverse of the transposed Jacobian of the system has rank < 6

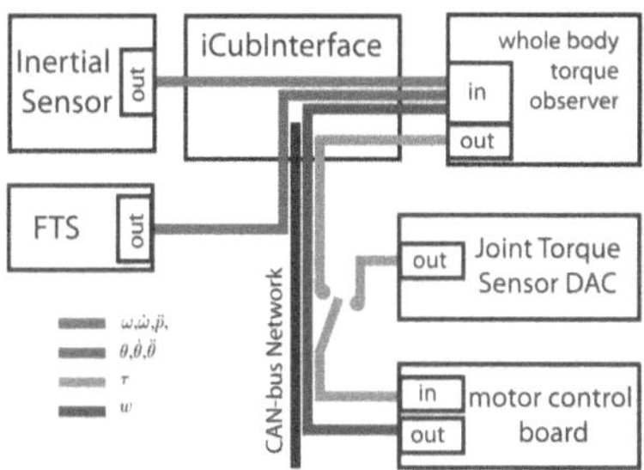

Fig. 6.10 Scheme of the software architecture that allow to perform the enrichment of the iCub haptic sensory system. The *wholeBodyTorqueObserver* module takes information from the inertial sensor and FTSs to perform the computation of joint torque to send to the motor control boards, that perform the control. The low level software architecture allows to chose the messages to read between the *virtual measurements* and actual measurements from joint level torque sensors

being ζ an arbitrary vector which span the Kernel of $J^\top$. It is noticeable here that $rank(J^\top) \in \mathbb{R}^{\leq 4}$, which means that $ker(J^\top) \in \mathbb{R}^{\neq 2}$. This means that there not exists a unique solution to w which satisfies $\tau = J^\top w$.

The Whole Body Torque Observer

The YARP module which performs the *virtual torque* estimation is called *WholeBodyTorqueObserver* [3]. This module uses the methods provided by the *iDyn* library (see the Doxygen documentation provided by [6]) in order to perform the computation of the *virtual joint torque measurements* which gives an estimation of the joint torques of 32 over 53 degrees of freedom of the iCub robot.

The iDyn library is a set of classes specifically designed within this work, which implements the methodology presented in Sect. 3, in order to estimate joint torques for interaction control, to calculate dynamic contribution for feedback linearization and, more in general, to perform inverse dynamic calculation. This library also allows inverse dynamic calculation performed over multiple branched chains, and it specifically provides classes for the iCub robot, as reported in Sect. 4.3.

The *wholeBodyTorqueObserver* module provides the estimation of joint torques (see Fig. 6.10). It uses data from the $3DOF$ orientation tracker placed on the iCub head, which is necessary to retrieve measurement of the absolute Cartesian velocity and acceleration (both linear and revolutionary) to calculate the other links kinematic quantities. It is also used to initialize the kinematic flow of information. The module also takes data from the four FTSs in order to perform wrench computation and therefore to compute the *virtual torque measurement*.

The estimated torques are sent to the *iCubInterface*, which provide specific ports dedicated to the *wholeBodyTorqueObserver* module for virtual torque data sharing.

Finally, the *iCubInterface* module send the information to the motor control boards, as presented previously in Sect. 6.3.2.

The Gravity Compensator

A second module, called *gravityCompensator*, have been implemented to improve the position control accuracy when interaction control is activated. This module performs an estimation of the gravitational component of the robot dynamic. The estimation is in fact used to reduce the steady state position error when the robot joints control modality is the impedance control mode, following the formulation presented in Sect. 5.2.3.

The data corresponding to the gravitational component of the robot dynamics is sent to the low level control boards as a torque reference. To compute this information, the module makes use of encoder data and of the absolute orientation of the robot position (which is obtained by the values of a $3DOF$ orientation tracker).

Considering the generic equation of a manipulation system:

$$\tau = M(\theta)\ddot{\theta} + \eta(\dot{\theta}, \theta) \tag{6.4}$$

where $M(\theta)$ is the mass matrix of the manipulator, and $\eta(\dot{\theta}, \theta)$ the vector of the other non linear terms (centrifugal, Corioli's and also gravitational components of the manipulator's dynamics). The gravitational contribution can be found from Eq. 6.4 by substituting $\dot{\theta} = 0$. The results gives an estimation of the gravitational components to be used as a joint torque reference to the torque controller implemented on the motor control boards, which follows the control law presented in Sects. 5.2.2, 5.2.3 and 5.2.4.

Therefore, the reference gravitational torque vector takes the form:

$$\tau_d = \tau_g = \hat{\eta}(0, \theta_j) \tag{6.5}$$

being $\hat{\eta}$ the estimation of the gravitational, centrifugal and Corioli's vector of Eq. 6.4.

Also for the gravitational terms, the *iCubInterface* provides a communication port which allows to set the reference torque to the motor control boards for feedback control. The software connection is presented in figure Fig. 6.11 where the yellow

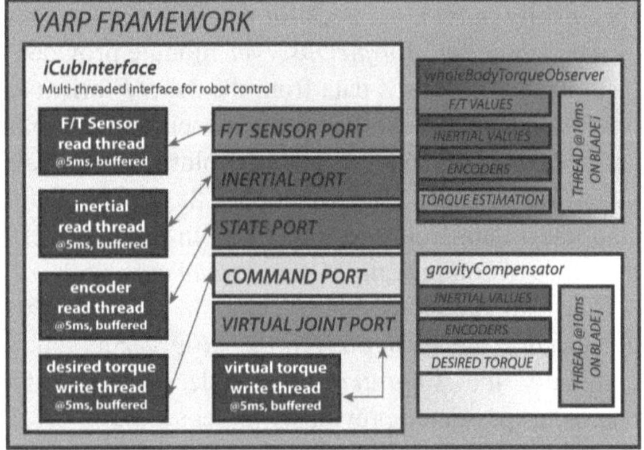

Fig. 6.11 Overall software modules that allow the enrichment of the information to the iCub robot

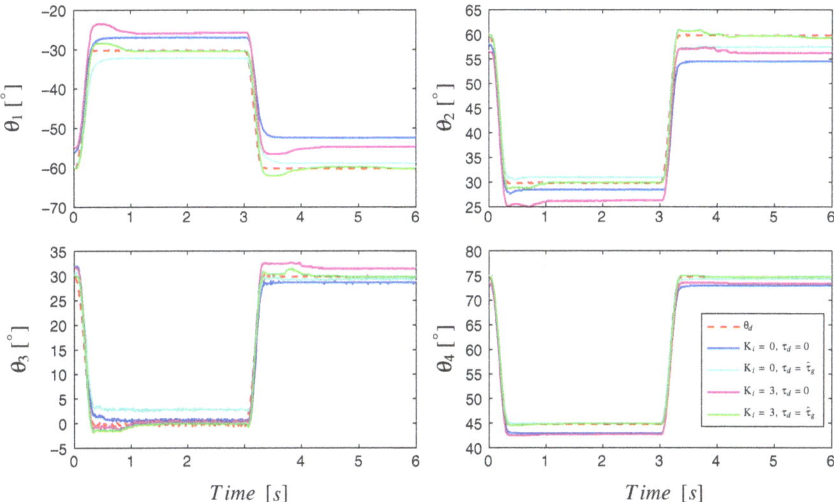

Fig. 6.12 The effect of the gravity compensation during a movement in *impedance mode*. It is also shown the effect of the integral component of the internal torque feedback on the steady state error

boxes are connected together by means of YARP ports. The module which performs the estimation of the gravitational component is a thread cycling with a rate of 10 ms.

Experimental results are shown in Fig. 6.12, where given a certain trajectory, the trajectory tracking of the system is presented with attention to the effect of the gravity compensation and also of the integral component of the torque regulator as shown in Sect. 5.2.4

References

1. Cannata, G., Maggiali, M., Metta, G., & Sandini, G. (2008). *An embedded artificial skin for humanoid robots*. in IEEE International Conference on Multisensor Fusion and Integration, Seoul, Korea.
2. Freescale. (2010). *CodeWarrior development tool*. http://www.freescale.com/webapp/sps/site/homepage.jsp?code=CW_HOME
3. Fumagalli, M. (2010). *iCub module computing the whole-body dynamics*. http://eris.liralab.it/iCub/main/dox/html/group__wholeBodyDynamics.html. GNU GPL v2.0
4. Fumagalli, M., Ivaldi, S., Randazzo, M., Natale, L., Metta, G., Sandini, G., et al. (2012). Force feedback exploiting tactile and proximal force/torque sensing—Theory and implementation on the humanoid robot iCub. *Auton. Robots*, *33*, 381–398.
5. Gamma, E., Helm, R., Johnson, R., & Vlissides, J. M. (1994). *Design patterns: Elements of reusable object-oriented software* (1st ed.). Reading: Addison-Wesley Professional.
6. Ivaldi, S., Fumagalli, M., & Pattacini, U. (2010). *Doxygen documentation of the iDyn library*. http://wiki.icub.org/iCub_documentation/idyn_introduction.html
7. Ivaldi, S., Fumagalli, M., Randazzo, M., Nori, F., Metta, G., & Sandini, G. (2011). *Computing robot internal/external wrenches by means of inertial, tactile and F/T sensors: Theory and*

implementation on the iCub. in 2011 11th IEEE-RAS International Conference on Humanoid Robots (Humanoids), (pp. 521–528).

8. http://eris.liralab.it. (2010). *The iCub user main page*. http://eris.liralab.it/wiki/Main_Page
9. Metta, G., Fitzpatrick, P., & Natale, L. (2006). Yarp: Yet another robot platform. *International Journal of Advanced Robotics Systems, special issue on Software Development and Integration in, Robotics*, 3.
10. Natale, L., Nori, F., Metta, G., Fumagalli, M., Ivaldi, S., Pattacini, U., et al. (2012). The iCub platform: A tool for studying intrinsically motivated learning. In G. Baldassarre & M. Mirolli (Eds.) *Intrinsically motivated learning in natural and artificial systems*. Berlin: Springer-Verlag.
11. Parmiggiani, A., Randazzo, M., Natale, L., Metta, G., & Sandini, G. (2009). *Joint torque sensing for the upper-body of the iCub humanoid robot*. In International Conference on Humanoid Robots, Paris, France.
12. Randazzo, M.,Fumagalli, M., Nori, F., Natale, L., Metta, G., & Sandini, G. (2011). *A comparison between joint level torque sensing and proximal F/T sensor torque estimation: Implementation on the iCub*. In 2011 IEEE/RSJ International Conference on Humanoid Robots (Humanoids), Intelligent Robots and Systems (IROS) (pp. 4161–4167).
13. Schmidt, D. C. (2003). *The adaptive communication environment*. http://www.cs.wustl.edu/schmidt/ACE.html
14. Schmidt, D. C., & Huston, D. H. (2002). *C++ network programming: Mastering complexity using ACE and patterns*. Verlag: Addison-Wesley Longman.

Appendix A
Improving the Estimate of Proprioceptive Measurements

In this respect, a possible algorithm for computing the better estimate of the kinematics, given the multiple sources, is briefly reported in Algorithm 3. Basically, given a set of K kinematics sources ▼, which for brevity we name $\kappa_1, \ldots, \kappa_K$, Algorithm 1 is solved K times. At each time k, κ_k is the only kinematic source which is not being removed from the EOG, and then the only ▼ in the graph. The solution of the EOG K times yields a set of conditional estimates $\omega_{j|\kappa_1}, \ldots, \omega_{j|\kappa_K}, \forall j$ (analogous considerations hold for $\dot{\omega}$ and $\ddot{p}$), which can be used by classical filters to provide the better estimate (e.g. maximum likelihood filters, Kalman filters *etc*). The analysis and evaluation of the possible filters has not been analyzed in this thesis, but future works might deal with this sort of estimation and improvement of internal robot perception.

Algorithm 3 Fusion of multiple kinematic sources

Require: EOG, $\kappa_k = [\omega_k, \dot{\omega}_k, \ddot{p}_k], k = 1, \ldots, K$
Ensure: $\hat{\omega}_i, \hat{\dot{\omega}}_i, \hat{\ddot{p}}_i, \forall i$
for all $k = 1 : K$ **do**
  Attach a node ▼ for κ_k
  Compute $\omega_{i|\kappa_k}, \forall i$
end for
Compute $\hat{\omega}_i$ = filter* $(\omega_{i|\kappa_1}, \ldots, \omega_{i|\kappa_K})$

* filter is a generic filter for data fusion from multiple sensors

M. Fumagalli, *Increasing Perceptual Skills of Robots Through Proximal Force/Torque Sensors*, Springer Theses, DOI: 10.1007/978-3-319-01122-6,

About the Author

Matteo Fumagalli was born on the 22nd of January 1982 in Rochester (Minnesota, USA). He studied mechanical engineer at Politecnico di Milano, Milan (Italy). He received the bachelor degree on April 2004 and the master degree on July 2006. After his graduation he worked for a short period (10/2006–12/2006) at the Department of Electronics and Computer Science of he Politecnico di Milano. On January 2007 he started the PhD student at the DIST—Department of Informatics, Systems and Telecomunication of the University of Genoa, Genoa (Italy). His PhD was in collaboration with the IIT—Fondazione Istituto Italiano di Tecnologia, Genoa (Italy), where he spent the four years of his PhD under the supervision of Dr. Francesco Nori.

After the graduation on April 2011, Mateo Fumagalli moved to the Netherlands where he is currently a post doctoral research fellow at the RAM—Robotics And Mechatronics group of the University of Twente, Enschede (NL). Here he conducts research on design, modeling and control of mechatronic systems for advanced robotic applications.

M. Fumagalli, *Increasing Perceptual Skills of Robots Through Proximal Force/Torque Sensors*, Springer Theses, DOI: 10.1007/978-3-319-01122-6,

Conclusions

During the last decade, trends in robotics foster research in the development of capabilities and skills which can make robots autonomous and safe (i.e. not dangerous). The human and robot coexistence in the same physical space require exploration, adaptation and learning of the autonomous system in order to create its knowledge of the effects of an action. The representation of the environment and the perception of the interaction are fundamental in this framework. Force information are of primary importance during the learning phase, as they become part of the experience of the autonomous system. Moreover, these information are necessary during the exploration process, also to prevent dangerous situation due to collision, through control.

In this thesis it has been shown a method which allows, through distributed (proximal) force sensors, inertial sensors and artificial skin, to increase the perceptual capabilities of the robot iCub. It has been shown that, under some assumptions, the method allows to have an estimation of the external force on any point of the robotic structure and, moreover, it allows to have an estimation of internal forces. These quantities have been used for control purposes. Impedance control and torque control at joint level has been implemented. Backdrivability performances have been risen through a model based approach that canceled the main source of dissipation of the iCub motors.

The software architecture and the modules that allow to perform excellent basic low level compliant behaviors has also been presented.

The lack of the artificial skin over the robotic structure did not allow to perform qualitative experiments of the dynamism of the method. This point will be achieved in near future works.

This thesis focused on the exploitation of force information with the goal of creating a framework for the exploration process. The work shown here is meant to supply for the necessity of increasing the perceptual capabilities of generic robotic systems for research in autonomous cognitive systems, but also for extending the representation of the generalized forces that arise in a physical human robot interaction scenario.

M. Fumagalli, *Increasing Perceptual Skills of Robots Through Proximal Force/Torque Sensors*, Springer Theses, DOI: 10.1007/978-3-319-01122-6,